Welcome

I Looked and This is What I Saw

This is a Thought Puzzle exploring

As Above, So Below

It tells a story, and each poster tells part of it.
Most of the story was told by Dr Rudolf Steiner
- till page 44 - where Glen Atkinson tells of how
Chemistry can be seen, within the Doctors Tale.

It presents a
Science of The Circle.

The posters are for Looking at
See what you know and
what you do not know

Follow the Thoughts and Questions
that arise; adding your Knowledge.

They are to be Lived With and Over Time,
The pieces of the Puzzle, fall into place and
a possible Universal Order will be revealed.

We are looking for What is Real,
There is nothing so Real as the Universe.

By looking at It, and Seeing its Order,
we can Understand our Environment,
as an expression of It, and Act along with It.

designed to be printed as A3 posters.

for more www.garudabd.org

ISBN 978-0-473-76551-4

Pdf ISBN 978-0-473-76552-1

CONTENTS

garuda@xtra.co.nz

The Apple of Life

Facing South

The energetic structure of all of creation

Vortexes on the vertical suck in matter and forces,
which are concentrated on the centre
and then squirted out along the horizontal plane

The circles are the magnetic field
created by this spinning movement

The small spirals within the vortexes
indicate the organisation of the six dimensions.

Painted by Chris Elliot

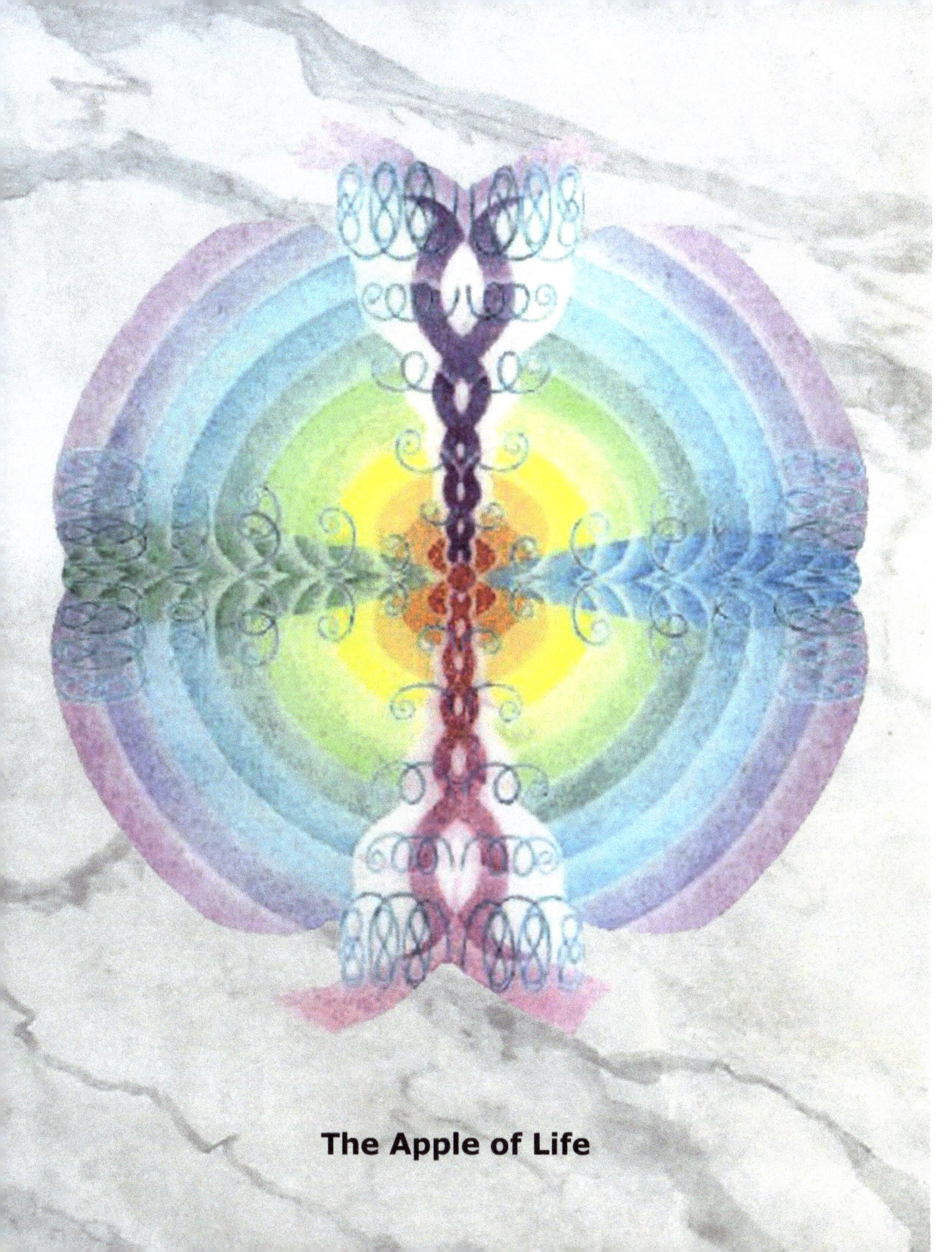

The Apple of Life

Gyroscope and Orientations

The apple of life is a 2 D representation
on what is really a gyroscopic 3D being

Observe the difference between
the north south axis
and the zenith and nadir axis

North South is the vertical
Zenith Nadir are the horizontal

The planets and the constellations beyond them
are all on the horizontal plane

Observe the 3 dimensions of
Up and Down
Forward and Backward
Left and Right

Observe how the organisation of the three planes
occur when they cross each other

Picture Orientations

Facing South

These diagrams are orientated
for the northern hemisphere,
facing south to the ecliptic .
A zenith orientation and
East is on the left hand side.

Magnetic North

Chemistry is electro magnetic
and orientates off magnetic north.
East is on the right hand.

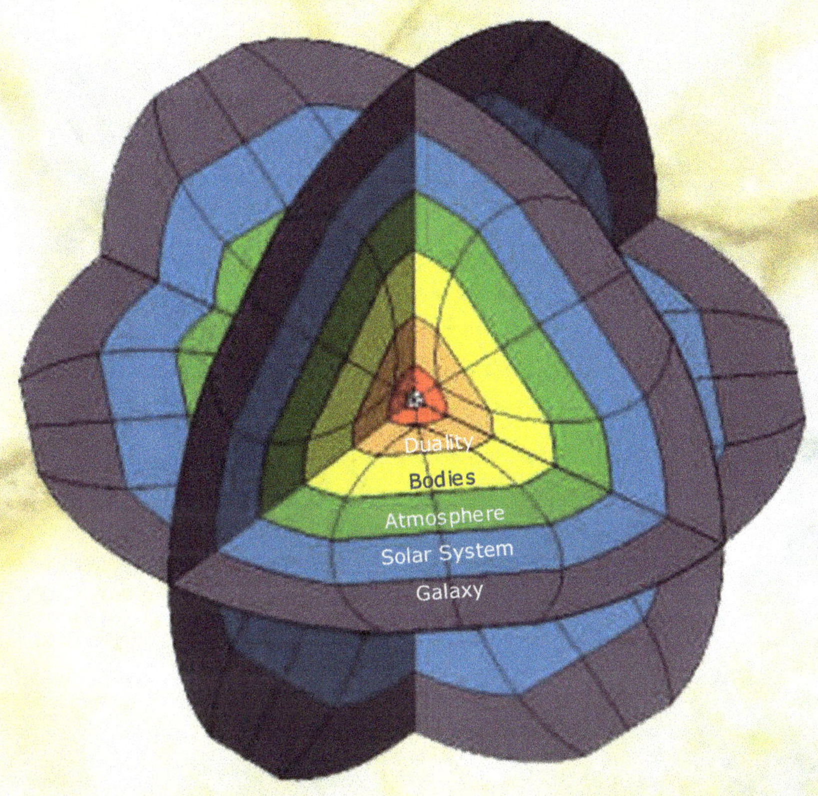

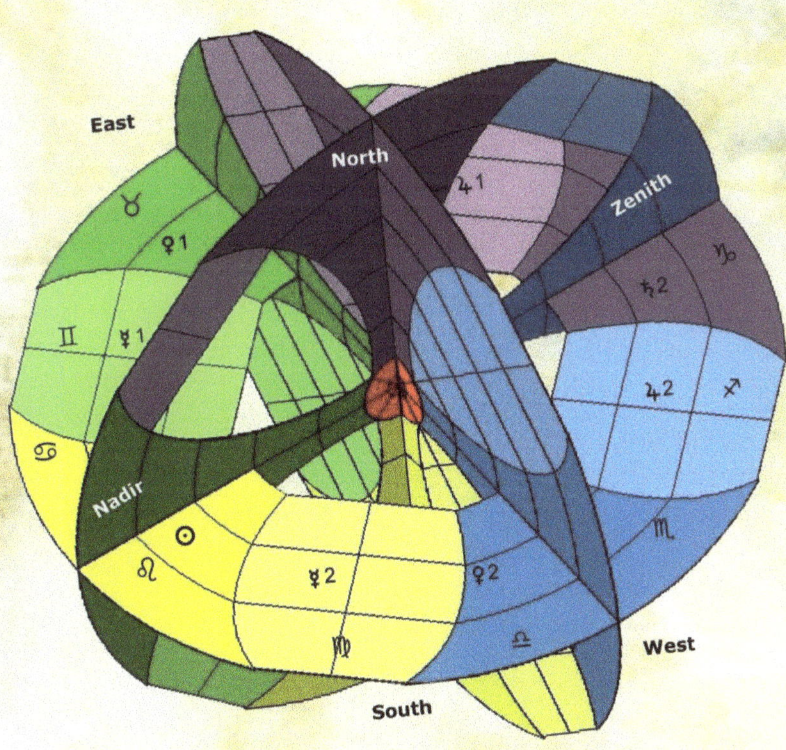

images by Sol Atkinson

The Three Worlds

The Three Organisations

Galaxy – Spirit World

Solar System – Soul World

Earth – Manifest World

All active in the same Time and Space

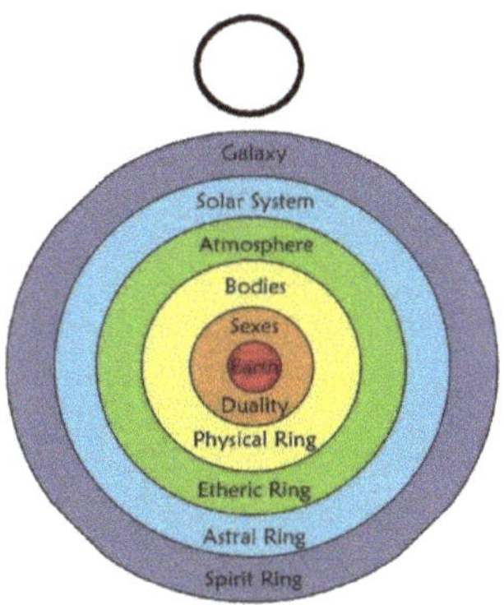

Stage 1
Archetype
Thesis
Passive
The Constant Field
Spirit
Saturn
Cosmic

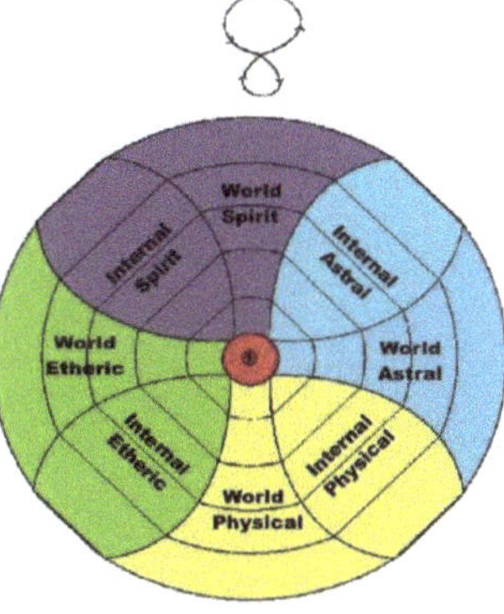

Stage 2
Forming
Antithesis
Pulsating
Surrounding the form
Soul
Jupiter
World and Internal

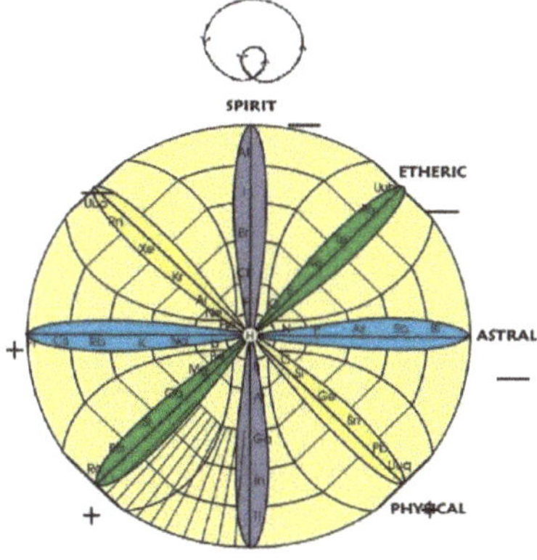

Stage 3
Manifestation
Synthesis
Enfolded
Form
Body
Chemistry

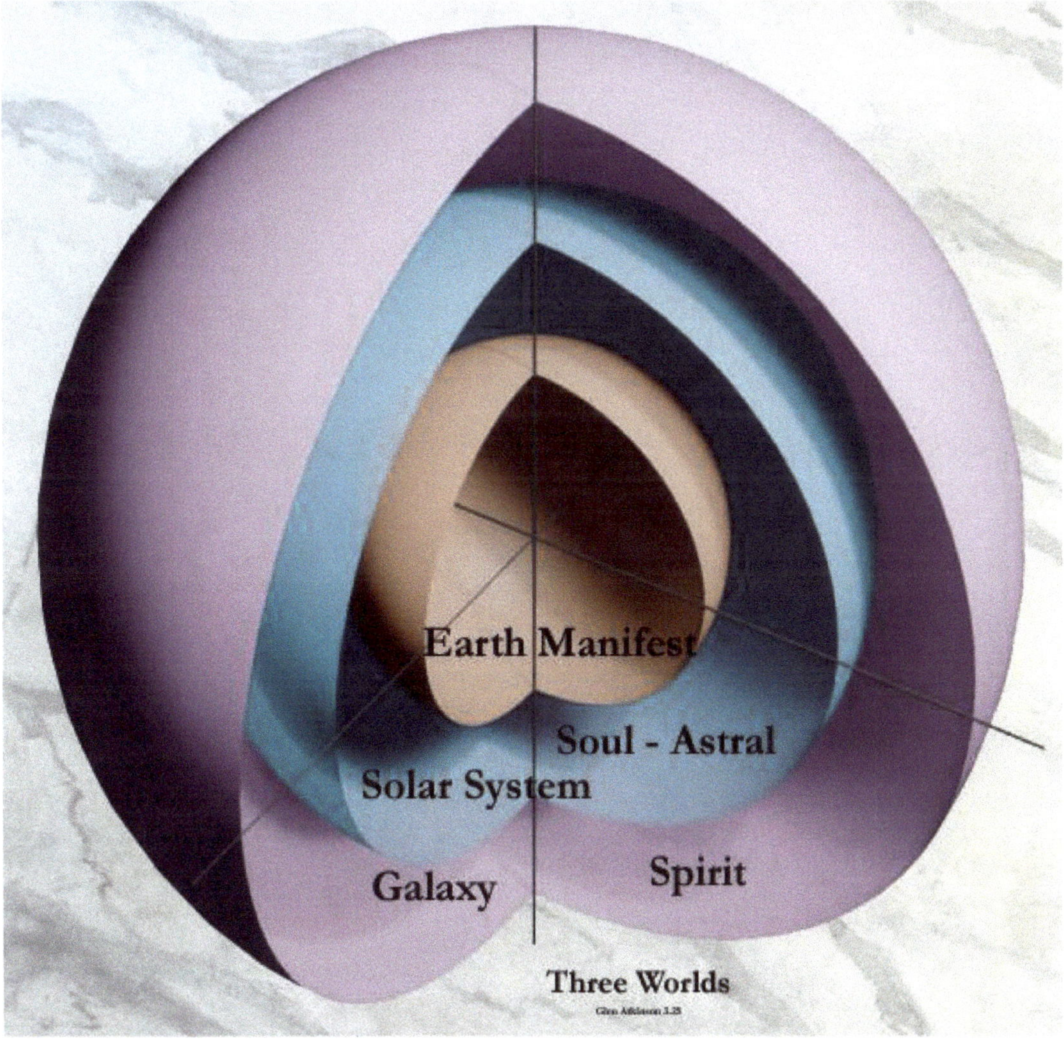

Earth Manifest
Soul - Astral
Solar System
Galaxy
Spirit

Three Worlds

The Four Worlds

Galaxy – Spirit World

Solar System – Soul World

Earth – Manifest World

And from the Earth came the

Atmosphere – Etheric World

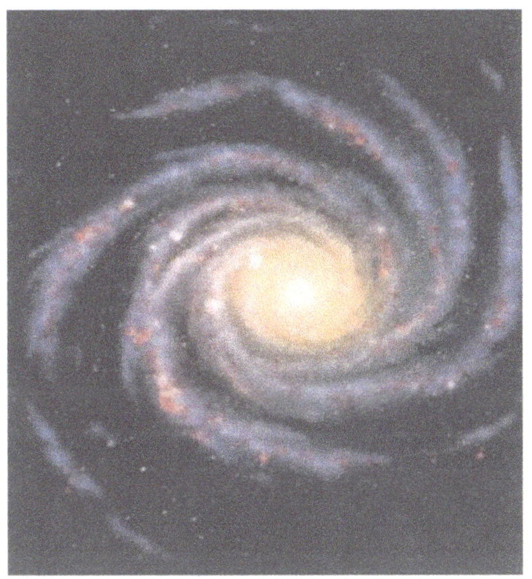

Galaxy

Spirit World

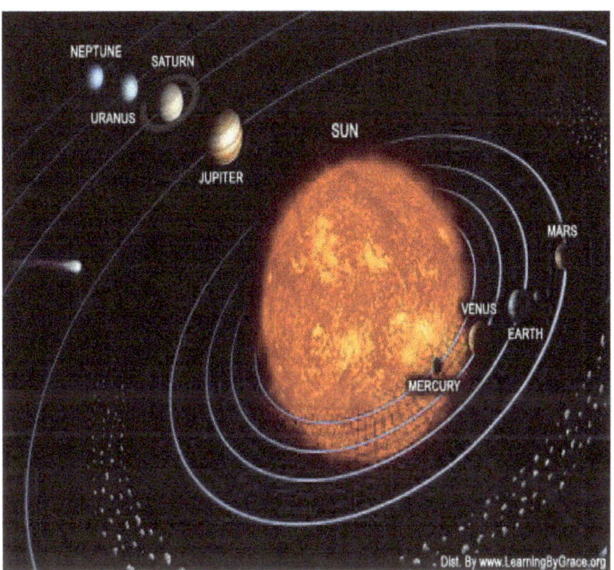

Solar System

Soul World

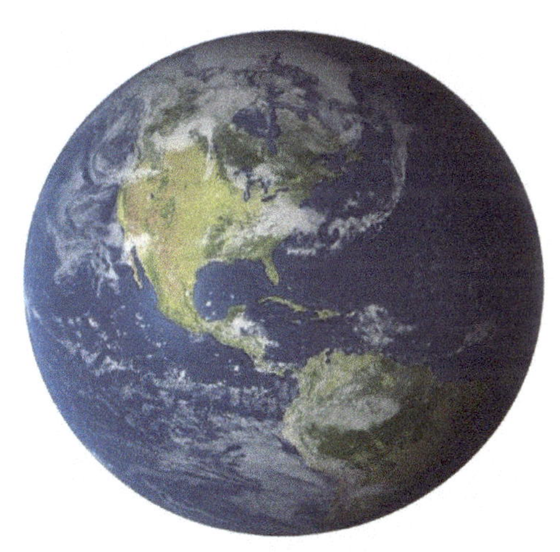

Earth

Manifest World

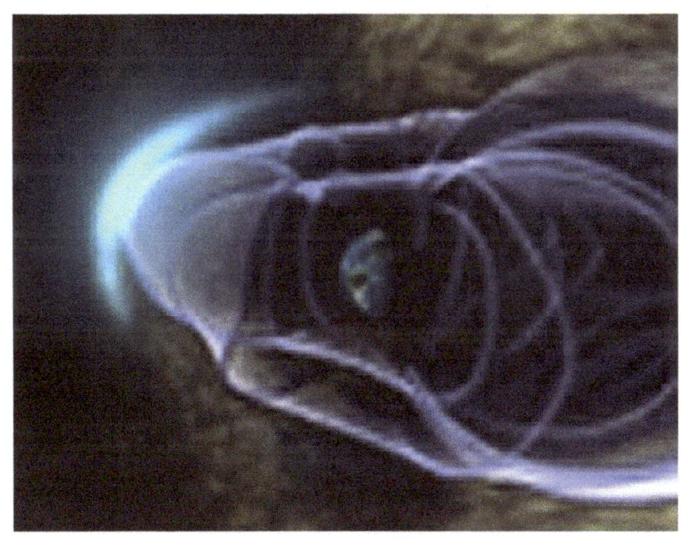

Atmosphere

Etheric World

The Layers of Creation

Facing South

The left spiral is the Astrological inward movement
through the six dimensions

Constellations, planets, elements,
modes duality, earth

The middle 'rings' diagram places this same information
as individual rings and associates them to
the energetic activities described by Dr Steiner.
This is our experience of creation.

The right gyroscope shows what happens
to the world activities once there is
movement that supports life forms.
The energetic activities polarise.

The world activities hold the vertical and horizontal axis
as the primary cross, while the internalised activities
of the lifeforms manifest as a secondary cross
in between the world activities.

The chart shows the 12 layers of creation

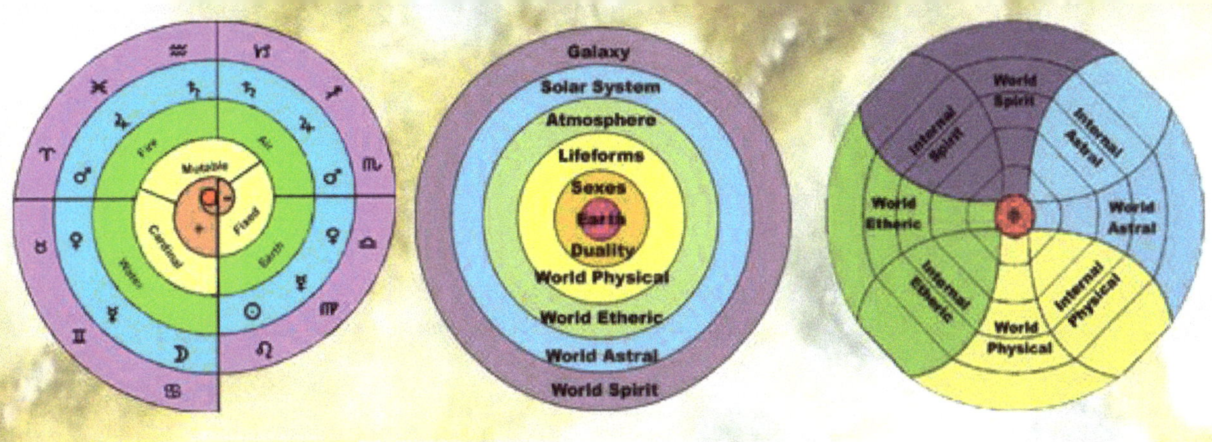

Layer		
SPIRIT Galaxy	♌ ♍ ♎ ♈ ♓ ♒	
	♋ ♊ ♉ ♏ ♐ ♑	*Spiritlands*
	☉ ↓ ⯘	*Collective Conscious*
	♅ ♆ ♇	*Collective Unconscious*
Solar System ASTRAL	♄ ♃ ♂	*Outer planets* Personal Conscious
	☽ ☿ ♀	*Inner planets* Personal Unconscio
ETHERIC Atmosphere	Warmth Light Chemical Life	*Ionosphere* Ethers
	Fire Air Water Earth	*Stratosphere* Elements
	Hydrogen Nitrogen Oxygen Carbon	*Troposphere*
	CHO Sugars Proteins 'Life'	*Biosphere*
PHYSICAL Earth	Mg. Na. K. Al. Mo Fe. Cu. Ag. Au. Pb	*Chemicals*
	Ca. Si.	*Solid Earth*

The layers of Creation

☽ Moon	♅ Uranus	♈ Aries	♎ Libra
☿ Mercury	♆ Neptune	♉ Taurus	♏ Scorpio
♀ Venus	♇ Pluto	♊ Gemini	♐ Sagittarius
♂ Mars	⯘ Persephone	♋ Cancer	♑ Capricorn
♃ Jupiter	↓ Vulcan	♌ Leo	♒ Aquarius
♄ Saturn	☉ Sun	♍ Virgo	♓ Pisces

A + B = C
of a Circle

Magnetic North

Creation = Movement + Time

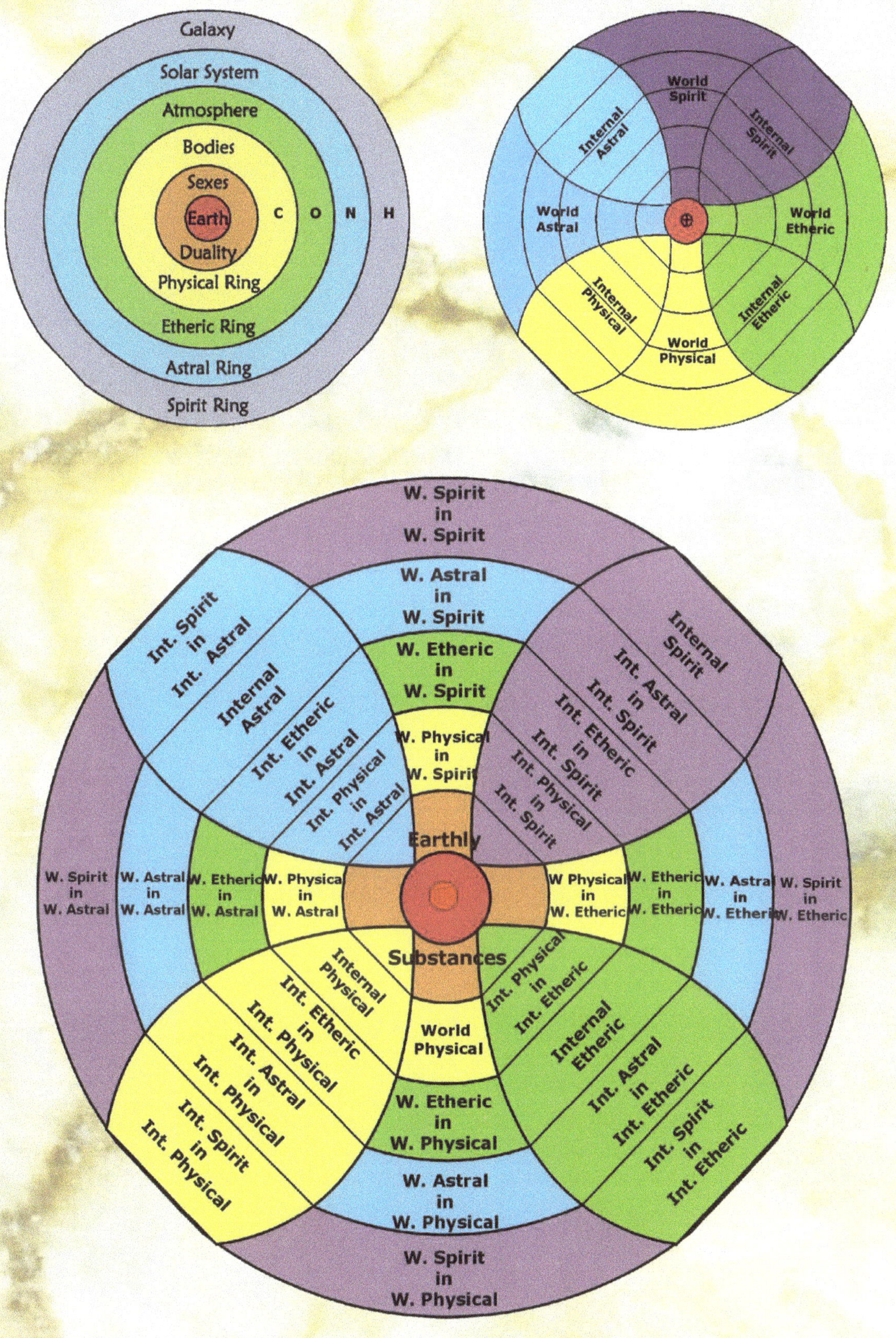

The A + B = C of the Circle

Glen Atkinson

The Kingdoms of Nature

All life are internalised manifestation
of the four energetic activities.

Observe the differences
between each kingdom.

Observe the polarity of
internalised and external activities

Both work all the time

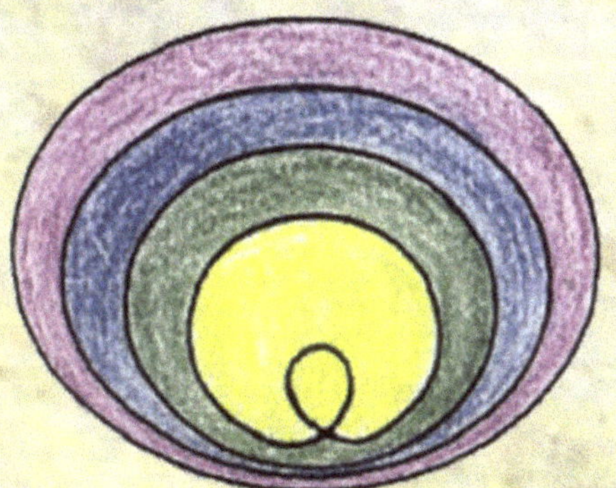

Mineral Kingdom
Internalised Physical Body

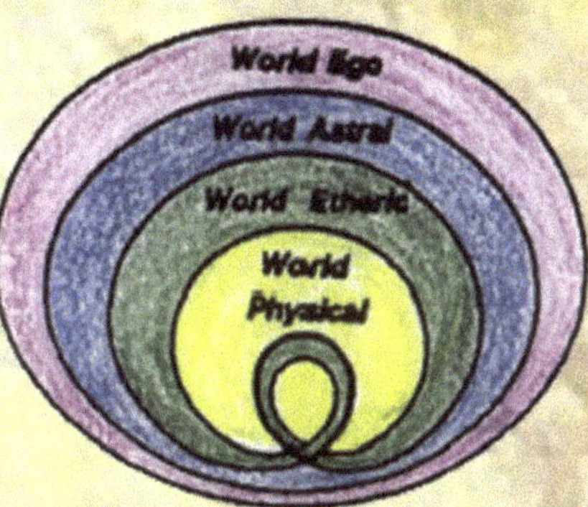

Plant Kingdom
Internalised Physical & Etheric Bodies

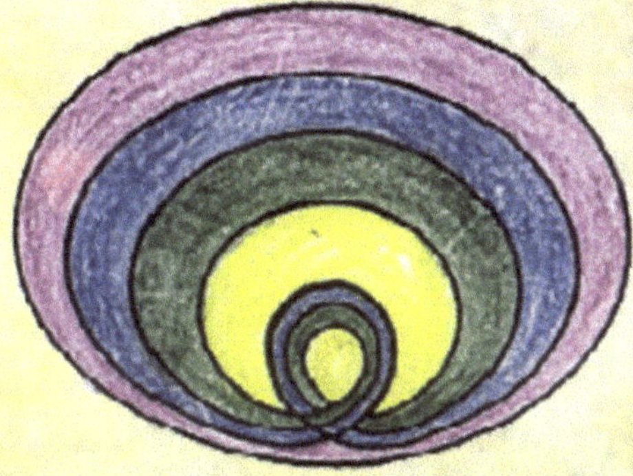

Animal Kingdom
Internalised Physical, Etheric & Astral Bodies

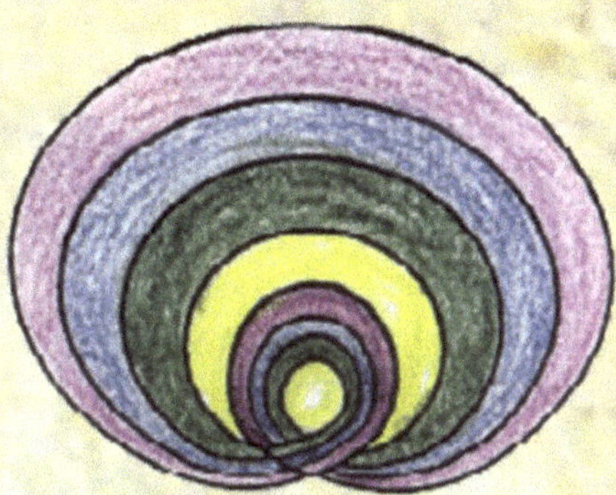

Human Kingdom
Internalised Physical, Etheric, Astral & Spirit Bodies

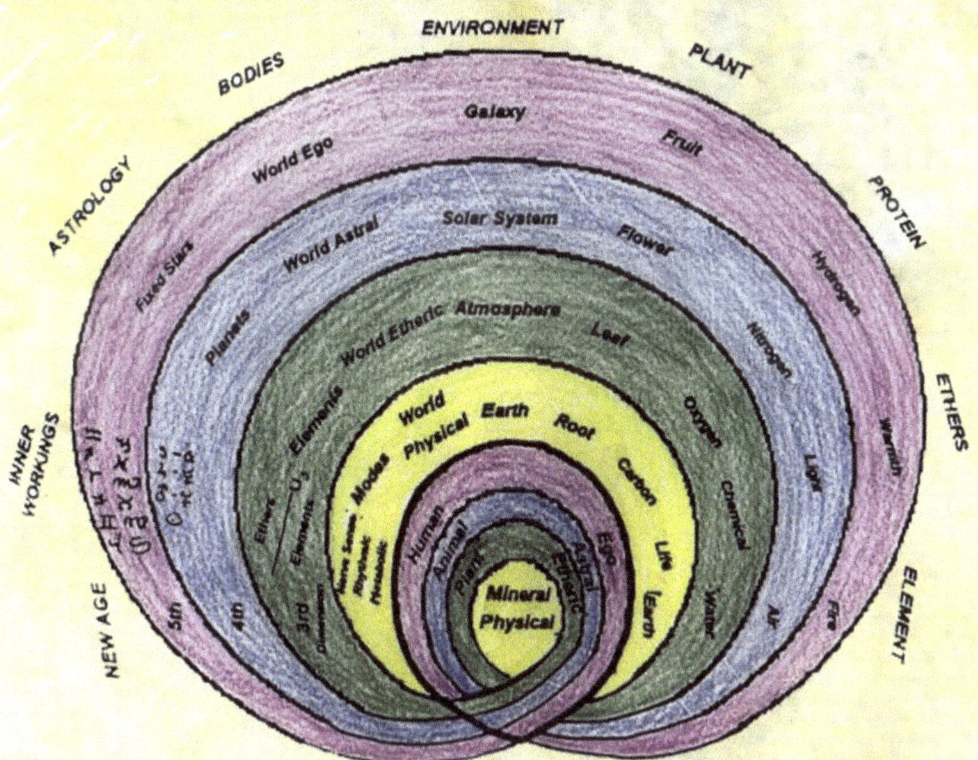

Spiritual Bodies and
the Kingdoms of Nature
1993

The ELECTRONIC BEING

Facing South

Cr = Mv + T

The Four Planes of Force
are interactive with each other

and Mediated by H N O C
in the Ring of Cosmic Elements

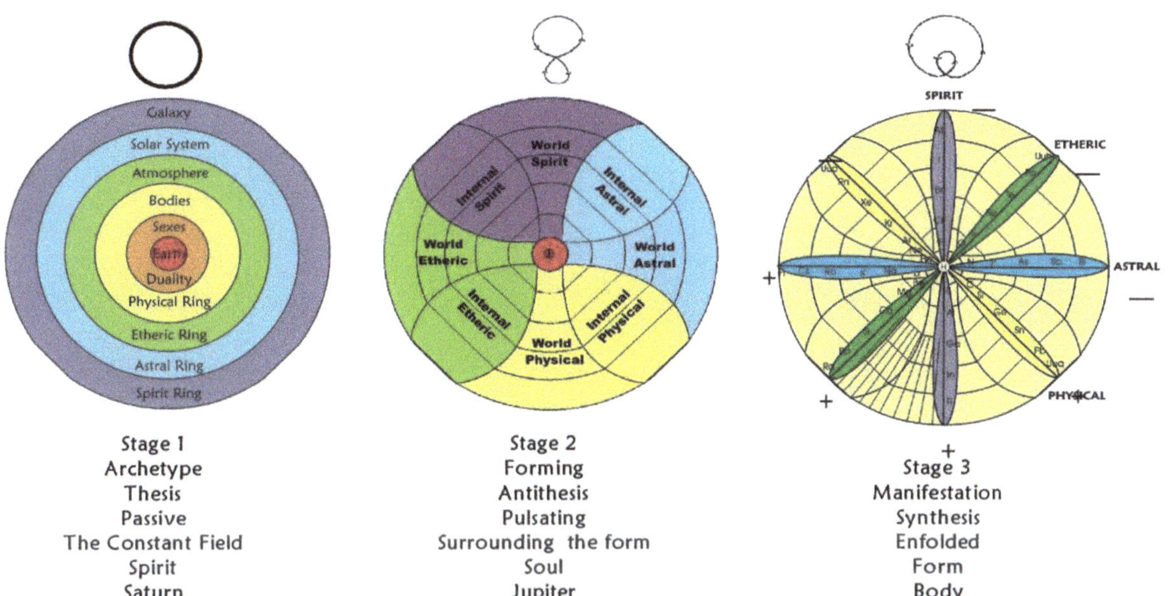

Stage 1	Stage 2	Stage 3
Archetype	Forming	Manifestation
Thesis	Antithesis	Synthesis
Passive	Pulsating	Enfolded
The Constant Field	Surrounding the form	Form
Spirit	Soul	Body
Saturn	Jupiter	Chemistry
Cosmic	World and Internal	

SPIRIT —
ETHERIC —
ASTRAL —
PHYSICAL +
MAGNETIC
DI ELECTRIC
ELECTRIC
ELECTRO-MAGNETIC
At I Br Cl F O N P As Sb Bi S Se Te Po Uuh Uuo Rn Xe Kr Ar Ne He H Fr Cs Rb K Na Li Be Mg Ca Sr Ba Ra C Si Ge Sn Pb Uuq Al Ga In Tl

The ELECTRONIC BEING

3D Electronic Being

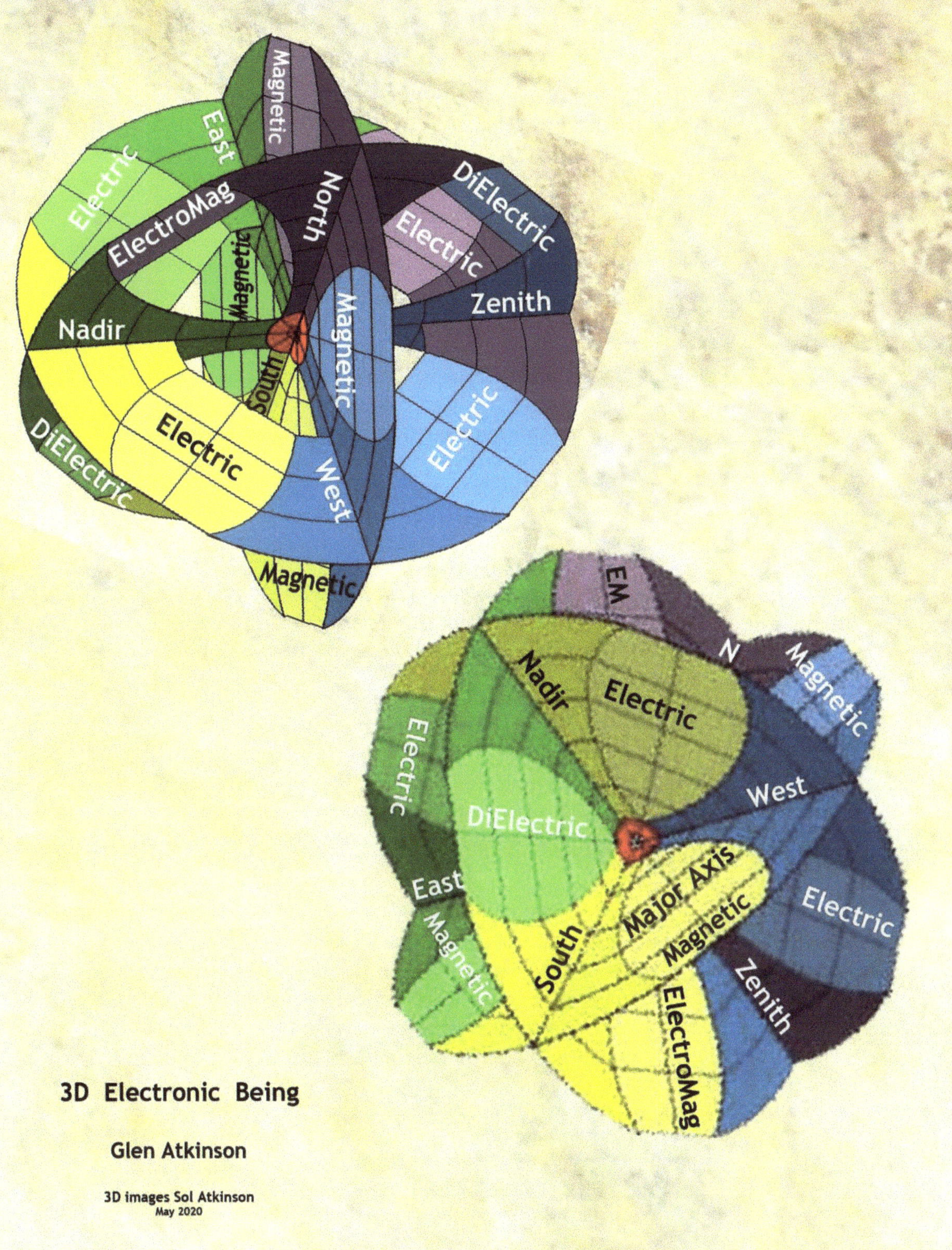

3D Electronic Being

Glen Atkinson

3D images Sol Atkinson
May 2020

" In every living organism
Physical, Ethereal, Astral and Spiritual forces are active.
In the plant, the *threefold* Physical and
the *fourfold* Ethereal forces, work from within outwards,
the Astral forces work around it,
and the Spiritual forces from the starry distance,
where the archetypes of the plants are formed"

"The effects of the Etheric Formative Forces are always general ones.
Never does an organism come into existence
merely through the Ethereal processes,
but only when an Astral principle
presses its seal upon the Ethereal"

" The working of the Astral principle
is carried into this Physical-Ethereal,
as a streaming, moving activity.
Manifesting archetypally in a sevenfoldness"

"The Spiritually working principles gather together the form
into a specific species and are archetypally arranged
according to the 12 principles, revealing themselves cosmically
in the Zodiac. It is only through this spiritual element that
'the plant' becomes a rose or a sage,
'the animal', a lion, and 'man a definite individuality."

"In the plant the seed is the carrier, or point of attachment,
of the Spiritual forces *which* form the species.
But the seed only unfolds when surrounded by
Physical, Ethereal and Astral elements in such a way
that these forces can stream freely together."

Dr. B Lievegoed 1951

'The Working of the Planets in the
Life Processes in Man and Earth'.

Biodynamics Decoded

CAPRICORN
Cardinal, Earth, Saturn 2
Limb Metabolic
Root Plants
eg Seeds of Carrots
Arthopoda - Insects
Aluminium

PISCES
Mutable, Water, Jupiter 1
Rhythmic
Leafing Plant
eg. Leaves of Cabbage
Mollusca - Shellfish
Chlorine

SCORPIO
Fixed, Water Mars 2
Nerve Sense
Leafing Plants
eg Roots of Cabbage
Amphibia - Frogs
Carbon

TAURUS
Fixed, Earth, Venus 1
Nerve Sense
Rooting Plants
eg Roots of Carrots
Echinodermata - Starfish
Nitrogen

VIRGO
Mutable, Earth, Mercury 2
Rhythmic
Rooting plants
eg Leaves of Carrots
Aves - Birds
Sodium

CANCER
Cardinal, Water, Moon 1
Limb Metabolic
Leaf plants
eg Fruit of Cabbages
Protozoa - Bacteria
Phosphorus

AQUARIUS
Fixed, Air, Saturn 1
Nerve Sense,
Flowering Plants
eg The roots of Roses
Vermes - Worms
Oxygen

SAGITTARIUS
Mutable, Fire, Jupiter 2
Rhythmic
Fruiting Plants
eg Leafs of the lemon tree
Pisces - Fishes
Magnesia

ARIES
Cardinal, Fire, Mars 1
Limb Metabolic
Fruiting Plants
eg Fruit of Peaches
Tunicata - Sea Squirts
Silica

LIBRA
Cardinal, Air, Venus 2
Limb Metabolic
Flower plants
eg Flowers of Roses
Reptilia - Lizards
Calcium

GEMINI
Mutable, Air, Mercury 1
Rhythmic
Flowering plants
eg Leaves of Roses
Coelentrata - Corals
Sulphur

LEO
Fixed, Fire, Sun 2
Nerve Sense,
Fruiting plants
eg Roots of Lemon trees
Mammals - Cows
Hydrogen

SATURN
Prep 507
Valerian tincture
Element - Phosphorus
Organ - Spleen
Metal - Lead
Strengthens Spirit

JUPITER
Prep 506
Dandelion/Mesentry
Element - Hydrogen
Organ - Liver
Metal - Tin
Helps Spirit and
Physical entwine

MARS
Prep 504
Nettle / Earth
Element - Nitrogen
Organ - Gall bladder
Metal - Iron
Harmonises the
Astral body

VENUS
Prep 502
Yarrow / Stag bladder
Element - Sulphur
Organ - Kidneys
Metal - Copper
Opens the Etheric
to the Astral

MERCURY
Prep 503
Chamomile /Intestines
Element - Oxygen
Organ - Intestines,
Metal-Mercury
Strengthens the
Etheric body

MOON
Prep 505
Oak Bark / Skull
Element - Carbon
Organ - Reproductive
Metal - Silver
Retards a
rampant Etheric

FIRE
Active Intuitive
Spirit - Will - Warmth ether
Seed & Fruit - Spheres
Adult, Hydrogen, Clay
Aries, Leo, Sagittarius
Cosmic Silica, Cosmic Forces

AIR
Rational Intellectual
Astral Body - Light ether
Flower - Pointed leaves
Pupa, Nitrogen, Sand
Gemini, Libra, Aquarius
Ter. Silica, Cosmic Matter

WATER
Nurturing, Emotional
Etheric Body - Chemical
Leaf & stems -Hemisphere
Grub, Oxygen, Humus
Cancer, Scorpio, Pisces
Cosmic Calcium, Earthly Forces

EARTH
Sustaining, Practical
Physical Body - Life ether
Root - Squares
Egg, Carbon, Lime
Taurus, Virgo, Capricorn
Ter. Calcium, Earthly Matter

FIXED
Consolidating - Antithesis
Nerve Sense system
The Head
Soil and Root
Salt - Basalt

MUTABLE
Changeable - Synthesis
Rhythmic system
Abdomen - Lungs & Heart
Leaf region
Mercury - Bentonite

CARDINAL
Initiating, Leading - Thesis
Limb Metabolic system
Digestion & Limbs
Flower & Fruit
Sulph - Sulphur

Galaxy
Solar System
Atmosphere
Lifeforms
Sexes
Earth
Duality
World Physical
World Etheric
World Astral
World Spirit

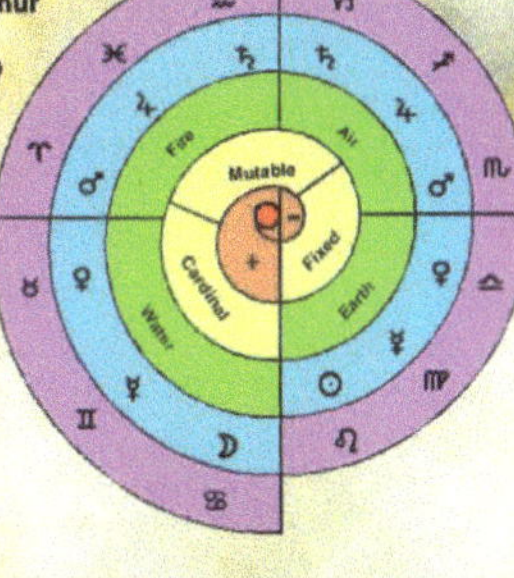

SUN +
Prep 501 - Silica / Horn
Spring & Summer
Dawn till Midday
Catabolic, Nutritive, Cold
Externally expansive and warm
Strengthens the upright growth
of the plant & enhances
the Astral & Spirit
Sun, Outer planets, Silica

MOON -
Prep 500 - Cow manure / Horn
Autumn & Winter
Afternoon & Evening
Anabolic, Reproductive, Warm
Externally contractive and cold
Enlivens the soil and growth
through the enhancing of the
Etheric and Physical bodies
Moon, Inner planets, Calcium

UNITY
Primal Oneness – Earth
This level is the manifest Earth
and the spiritual state of unity
from which all life begins and
returns. The atom acts
as the doorway between
spirit and matter

The references on this diagram
are drawn from the work of
Dr R Steiner, Dr R Hauschka
Dr E Kolisko, Dr Lievegeod
& Glen Atkinson

World Spirit
Astral FF
Free Fire
Cosmic Silica
Cosmic Matter
Bound Light
World Etheric
Free Water
Cosmic Calcium
Earth
Earthly Silica
Free Light
World Astral
Earthly Calcium
Earthly Matter
Bound Life Ether
Free Earth
Etheric FF
Physical FF
World Physical

www.garudabd.org

Gyroscopic Agriculture

Facing South

How do the activities work into each other?

Put diagram 3 on top of diagram 4
To get the bottom left picture

The picture on the bottom right
are the words RS used to
describe these activities

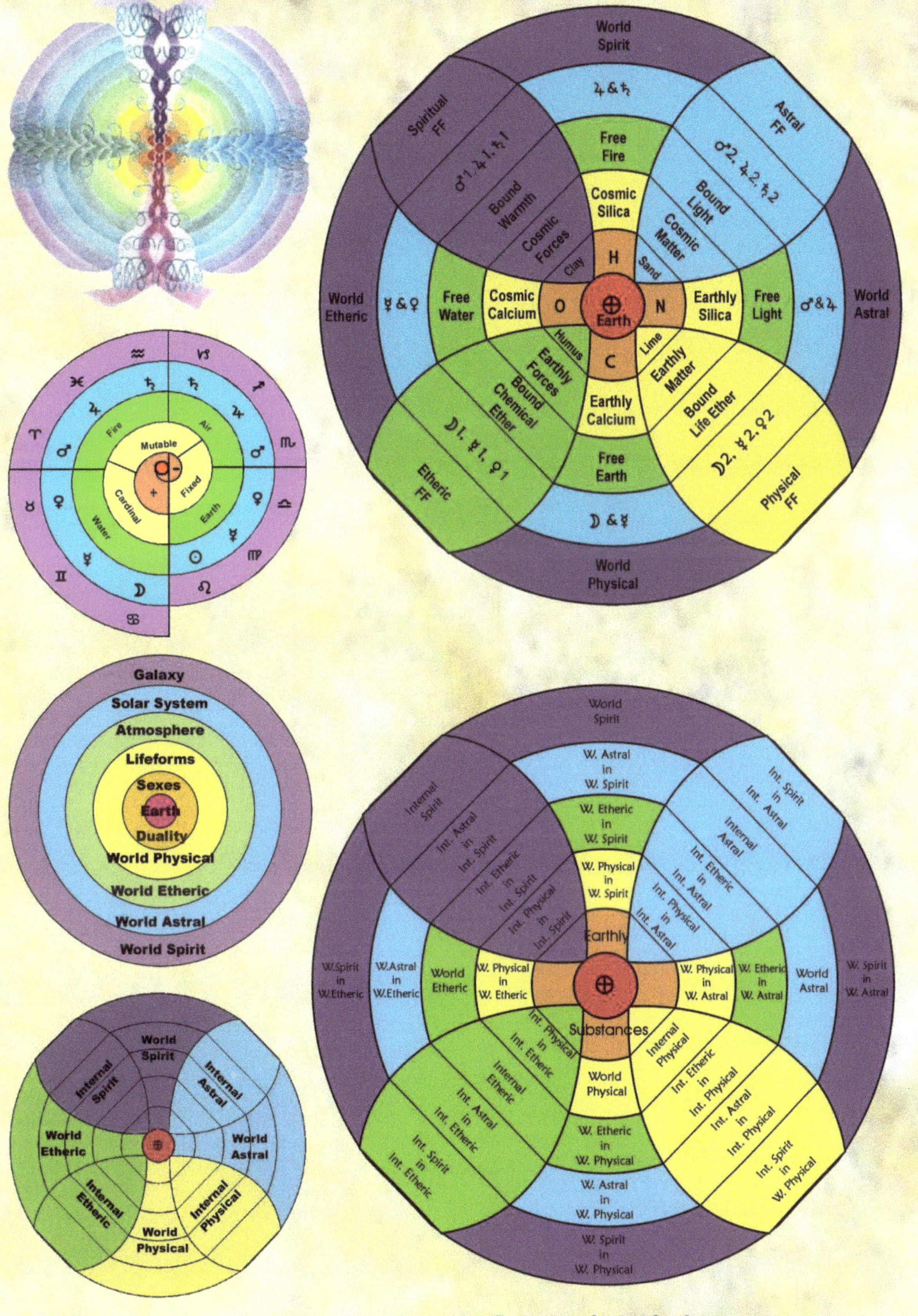

Mutable
Fire
Air
Cardinal
Fixed
Water
Earth
Galaxy
Solar System
Atmosphere
Lifeforms
Sexes
Earth
Duality
World Physical
World Etheric
World Astral
World Spirit
World Spirit
Internal Spirit
Internal Astral
World Etheric
World Astral
Internal Etheric
Internal Physical
World Physical
World Spirit
Spiritual FF
Astral FF
Free Fire
Cosmic Silica
Bound Warmth
Cosmic Forces
Bound Light
Cosmic Matter
Clay
Sand
H
World Etheric
Free Water
Cosmic Calcium
O
Earth
N
Earthly Silica
Free Light
World Astral
Humus
Lime
C
Earthly Forces
Bound Chemical Ether
Earthly Matter
Bound Life Ether
Earthly Calcium
Free Earth
Etheric FF
Physical FF
World Physical
World Spirit
W. Astral in W. Spirit
Internal Spirit
Int. Spirit in Int. Astral
W. Etheric in W. Spirit
Int. Astral in Int. Spirit
Internal Astral
W. Physical in W. Spirit
Int. Etheric in Int. Spirit
Int. Etheric in Int. Astral
Int. Physical in Int. Spirit
Int. Physical in Int. Astral
Earthly
W.Spirit in W.Etheric
W.Astral in W.Etheric
World Etheric
W. Physical in W. Etheric
W. Physical in W. Astral
W. Etheric in W. Astral
World Astral
W. Spirit in W. Astral
Int. Physical in Int. Etheric
Substances
Internal Physical
Internal Etheric
World Physical
Int. Etheric in Int. Physical
Int. Astral in Int. Etheric
Int. Astral in Int. Physical
W. Etheric in W. Physical
Int. Spirit in Int. Etheric
Int. Spirit in W. Physical
W. Astral in W. Physical
W. Spirit in W. Physical
Gyroscopic Agriculture
Glen Atkinson 8.8.2008

Creation = Movement + Time

Magnetic North

It moves through three stages

Archetype

Behind Manifestation

Manifestation

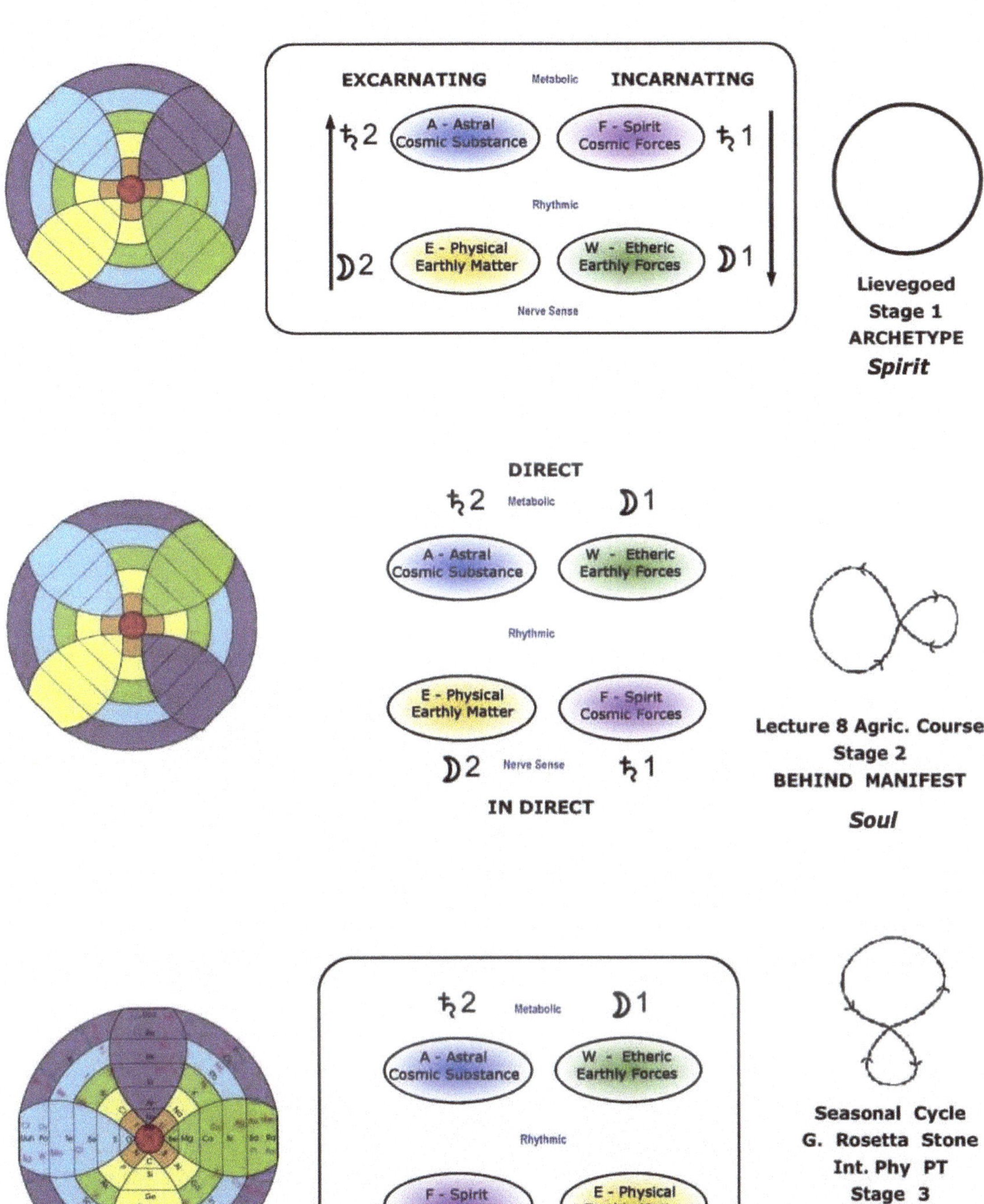

The Planets in the Three Worlds

Glen Atkinson

Planets in the Steiner's Stories

Magnetic North

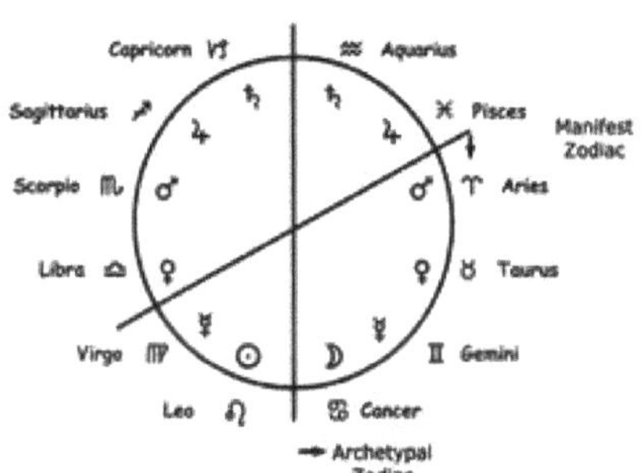

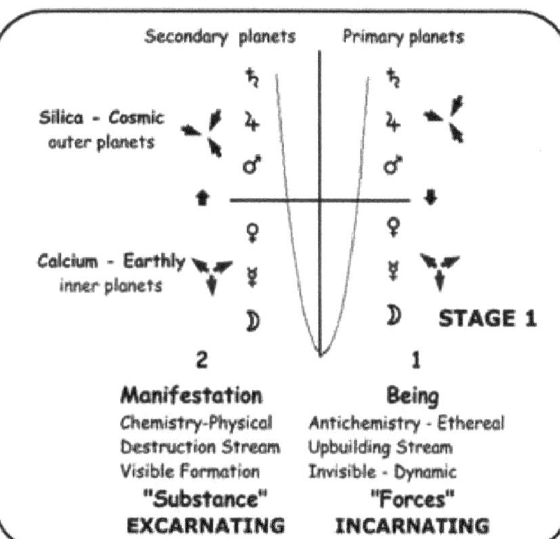

Lievegoed
Spirit

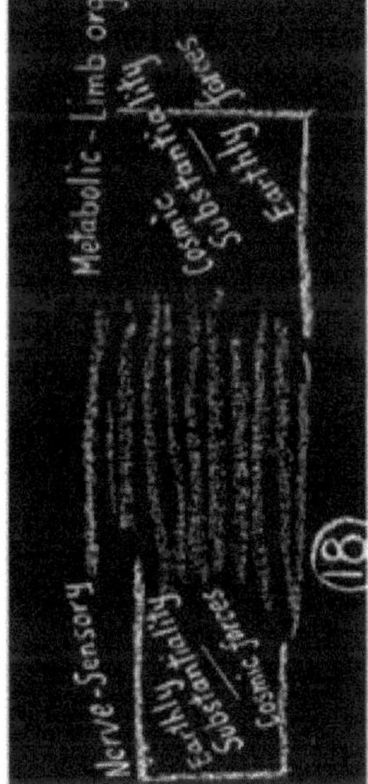

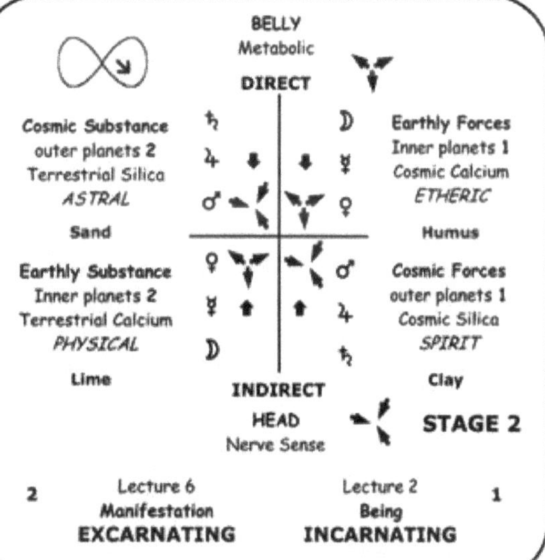

Steiners Agriculture Course
Soul

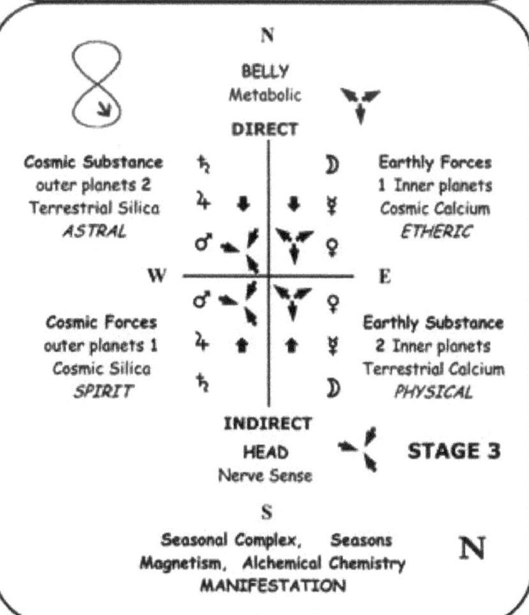

When we spin through the Seasons In the Saps and Ethers
Physical

The Planets in the Three Worlds 2

Glen Atkinson

9.1.21

Ag Course 2

Facing South

The second story of the agriculture course tells of how the activities work INSIDE living beings.

Things polarise

So we are taken one step towards manifestation
From the Spirit World, through a lemniscate twist,
into the Soul World

Where

The internal etheric and internal spirit
change places

And the physical organism is sustained
through the Physical Formative Forces

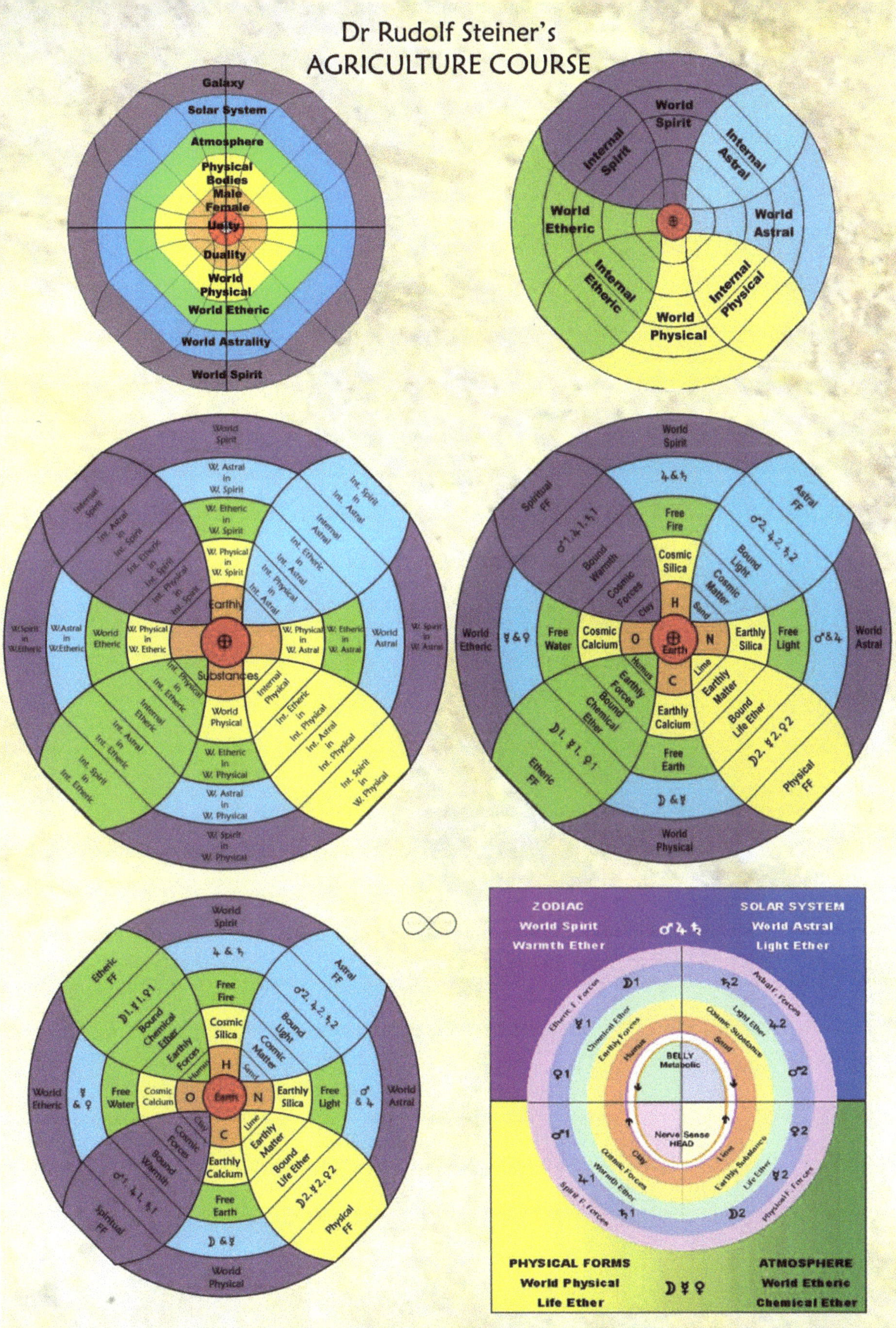

Dr Rudolf Steiner's
AGRICULTURE COURSE
Galaxy
Solar System
Atmosphere
Physical Bodies
Male
Female
Unity
Duality
World Physical
World Etheric
World Astrality
World Spirit
World Spirit
Internal Spirit
Internal Astral
World Etheric
World Astral
Internal Etheric
Internal Physical
World Physical
World Spirit
W. Astral in W. Spirit
W. Etheric in W. Spirit
W. Physical in W. Spirit
Internal Spirit
Int. Astral in Int. Spirit
Int. Etheric in Int. Spirit
Int. Physical in Int. Spirit
Int. Spirit in Int. Astral
Internal Astral
Int. Etheric in Int. Astral
Int. Physical in Int. Astral
Earthly
Substances
W.Spirit in W.Etheric
W.Astral in W.Etheric
World Etheric
W. Physical in W. Etheric
W. Physical in W. Astral
W. Etheric in W. Astral
World Astral
W. Spirit in W. Astral
Int. Physical in Int. Etheric
Internal Etheric
Int. Astral in Int. Etheric
Int. Spirit in Int. Etheric
Internal Physical
Int. Etheric in Int. Physical
Int. Astral in Int. Physical
Int. Spirit in W. Physical
World Physical
W. Etheric in W. Physical
W. Astral in W. Physical
W. Spirit in W. Physical
World Spirit
♃ & ♄
Spiritual FF
♂1, ♃1, ♄1
Bound Warmth
Cosmic Forces
Free Fire
Cosmic Silica
Astral FF
♂2, ♃2, ♄2
Bound Light
Cosmic Matter
Clay
Sand
H
O
N
C
Earth
World Etheric
☿ & ♀
Free Water
Cosmic Calcium
Earthly Silica
Free Light
♂ & ♃
World Astral
Humus
Lime
Earthly Forces
Bound Chemical Ether
Earthly Matter
Bound Life Ether
Earthly Calcium
☽1, ☿1, ♀1
☽2, ☿2, ♀2
Etheric FF
Free Earth
Physical FF
☽ & ☿
World Physical
World Spirit
♃ & ♄
Etheric FF
☽1, ☿1, ♀1
Bound Chemical Ether
Earthly Forces
Free Fire
Cosmic Silica
Astral FF
♂2, ♃2, ♄2
Bound Light
Cosmic Matter
Humus
Sand
H
O
N
C
Earth
World Etheric
☿ & ♀
Free Water
Cosmic Calcium
Earthly Silica
Free Light
♂ & ♃
World Astral
Clay
Lime
Cosmic Forces
Bound Warmth
Earthly Matter
Bound Life Ether
Earthly Calcium
♂1, ♃1, ♄1
☽2, ☿2, ♀2
Spiritual FF
Free Earth
Physical FF
☽ & ☿
World Physical
ZODIAC
World Spirit
Warmth Ether
♂ ♃ ♄
SOLAR SYSTEM
World Astral
Light Ether
Etheric F. Forces
Chemical Ether
Earthly Forces
Humus
Astral F. Forces
Light Ether
Cosmic Substance
Sand
BELLY
Metabolic
Nerve Sense
HEAD
Cosmic Forces
Clay
Warmth Ether
Spirit F. Forces
Lime
Earthly Substance
Life Ether
Physical F. Forces
☽1
☿1
♀1
♂1
♃1
♄1
♄2
♃2
♂2
♀2
☿2
☽2
PHYSICAL FORMS
World Physical
Life Ether
☽ ☿ ♀
ATMOSPHERE
World Etheric
Chemical Ether

3x4

Magnetic North

The internal activities
within the energetic bodies

The organisation of Soul World
in Nature

The internal activities take place
within the external realities

Stage 2

nerve-sense syst.
metabolic-limb syst.
earthly substance
cosmic substance
cosmic forces
earthly forces
2
1

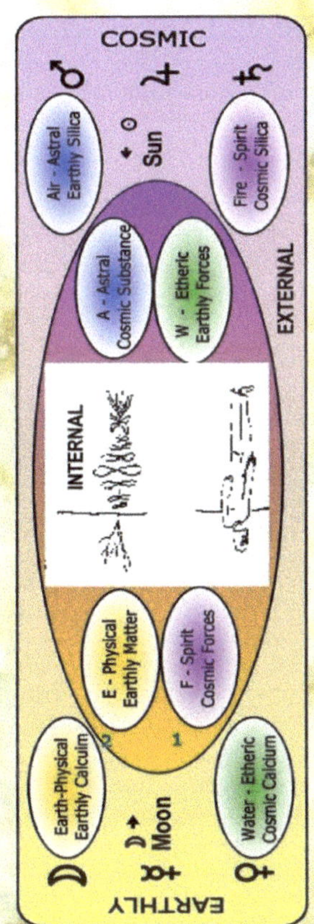

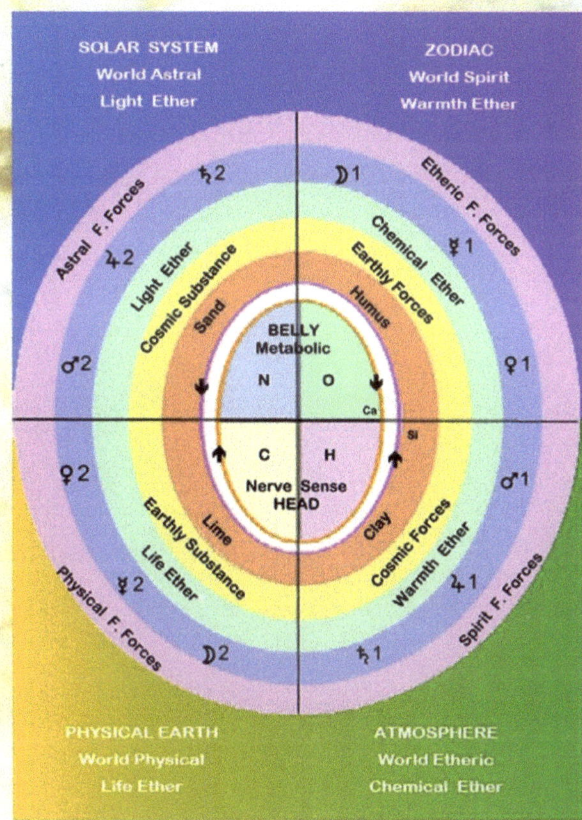

The Organisation of Dr Steiner's Agriculture Course

Lecture 2 & 8 - Magnetic North

Glen Atkinson

Fourfold Harmonies

Facing South

The Four activities manifest
at every layer of manifestation

Can you add some?

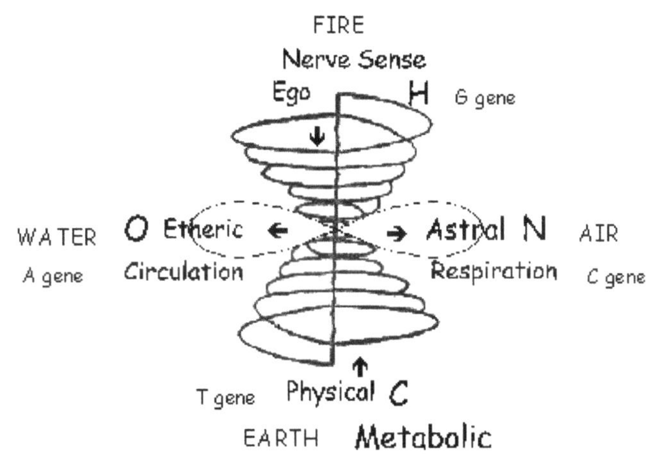

Galaxy	Solar System	Atmosphere	Earth	*Spheres*
Spirit	Astral	Etheric	Physical	*Bodies*
Human	Animal	Plant	Mineral	*Kingdoms*
Will	Psychology	Immunity	Body	*Human*
Nerve Sense	Respiratory	Circulation	Metabolic	*Body Systems*
Warmth	Light	Chemical	Life	*Ethers*
Fire	Air	Water	Earth	*Elements*
Cos. Forces	Cos Substance	Ear. Forces	Ear. Substance	*Phy F F*
Beetle	Pupa / Butterfly	Grub	Egg	*Insect*
Seed	Ripeness	Size	Quality	*Fruit*
Germ	Seed Coat	Cotyledons	Viability	*Seed*
Fruit &Seed	Flower	Leaf	Root	*Plant*
Roundness	Pointed	Wavey	Square	*Forms*
Stalk	Skin	Mass	Tissues	*Plant Growth*
Nucleus	Mitochondria	Cytoplasm	Cell Tissues	*Cell*
Hydrogen	Nitrogen	Oxygen	Carbon	*Biochemistry*
G	C	A	T	*DNA*
Clay	Sand	Humus	Lime	*Soil*
Cos. Silica	Ear. Silica	Cos. Calcium	Ear. Calcium	*Ca & Si*
North	West	East	South	*Magnetic*

Stage 3

Magnetic North

Manifestation

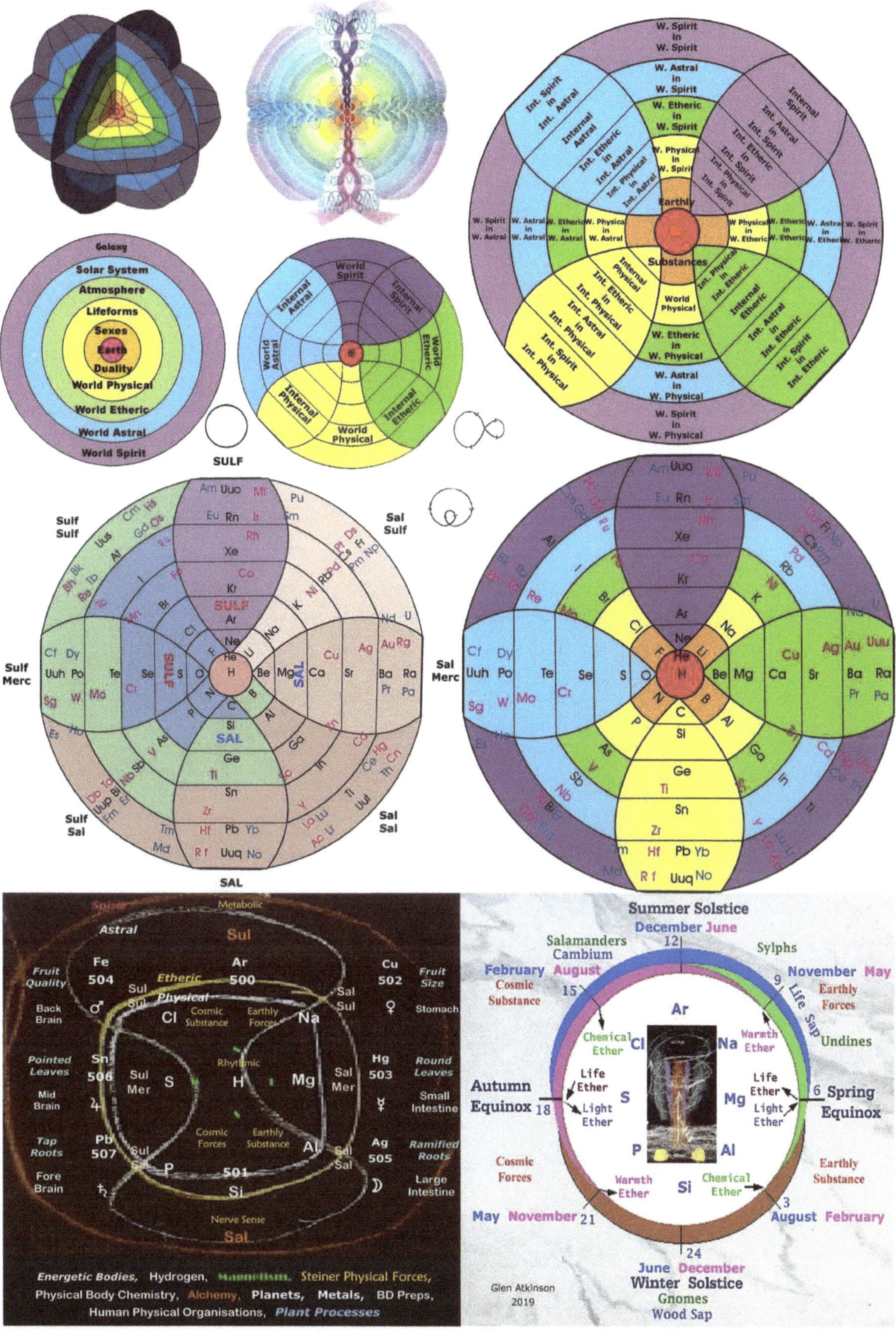

W. Spirit in W. Spirit
W. Astral in W. Spirit
W. Etheric in W. Spirit
W. Physical in W. Spirit
Int. Spirit in Int. Astral
Internal Astral
Int. Etheric in Int. Astral
Int. Physical in Int. Astral
Internal Spirit
Int. Astral in Int. Spirit
Int. Etheric in Int. Spirit
Int. Physical in Int. Spirit
Earthly
Substances
W. Spirit in W. Astral
W. Astral in W. Astral
W. Etheric in W. Astral
W. Physical in W. Astral
W Physical in W. Etheric
W. Etheric in W. Etheric
W. Astral in W. Etheric
W. Spirit in W. Etheric
Internal Physical
Int. Etheric in Int. Physical
Int. Astral in Int. Physical
Int. Spirit in Int. Physical
World Physical
Int. Physical in Int. Etheric
Internal Etheric
Int. Astral in Int. Etheric
Int. Spirit in Int. Etheric
W. Etheric in W. Physical
W. Astral in W. Physical
W. Spirit in W. Physical
Galaxy
Solar System
Atmosphere
Lifeforms
Sexes
Earth
Duality
World Physical
World Etheric
World Astral
World Spirit
World Spirit
Internal Spirit
Internal Astral
World Astral
World Etheric
Internal Physical
World Physical
Internal Etheric
SULF
Sulf Sulf
Sal Sulf
Sulf Merc
Sal Merc
Sulf Sal
Sal Sal
SAL
SULF
SAL
Spirit
Metabolic
Astral
Sul
Fruit Quality
Fe 504
Ar 500
Cu 502
Fruit Size
Etheric
Physical
Sul Sul
Sal Sul
Back Brain
Cl
Cosmic Substance
Earthly Forces
Na
Stomach
Pointed Leaves
Sn 506
Rhythmic
Hg 503
Round Leaves
Sul Mer
S
H
Mg
Sal Mer
Mid Brain
Small Intestine
Cosmic Forces
Earthly Substance
Tap Roots
Pb 507
Al
Ag 505
Ramified Roots
Sul Sal
P
501
Sal Sal
Fore Brain
Si
Large Intestine
Nerve Sense
Sal
Energetic Bodies, Hydrogen, Magnetism, Steiner Physical Forces,
Physical Body Chemistry, Alchemy, Planets, Metals, BD Preps,
Human Physical Organisations, Plant Processes
Summer Solstice
December June
12
Salamanders Cambium
Sylphs
February August
15
Cosmic Substance
November May
9
Earthly Forces
Life Sap
Undines
Ar
Chemical Ether
Cl
Na
Warmth Ether
Autumn Equinox
18
Life Ether
Light Ether
S
Mg
Life Ether
Light Ether
6
Spring Equinox
P
Al
Cosmic Forces
Earthly Substance
Warmth Ether
Si
Chemical Ether
May November 21
3
August February
24
June December
Winter Solstice
Gnomes
Wood Sap
Glen Atkinson
2019

Glenological Rosetta Stone

Magnetic North

Biodynamics

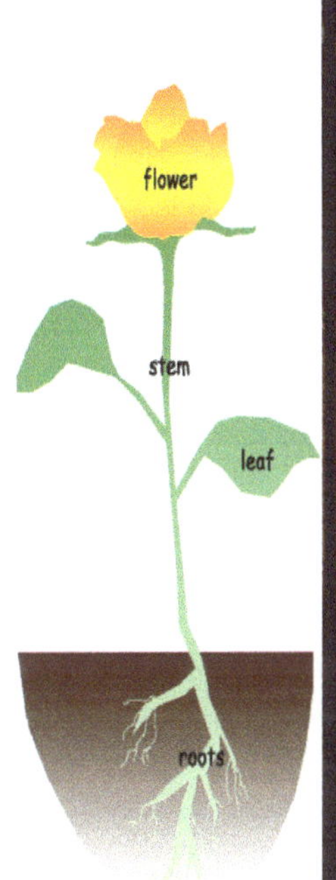

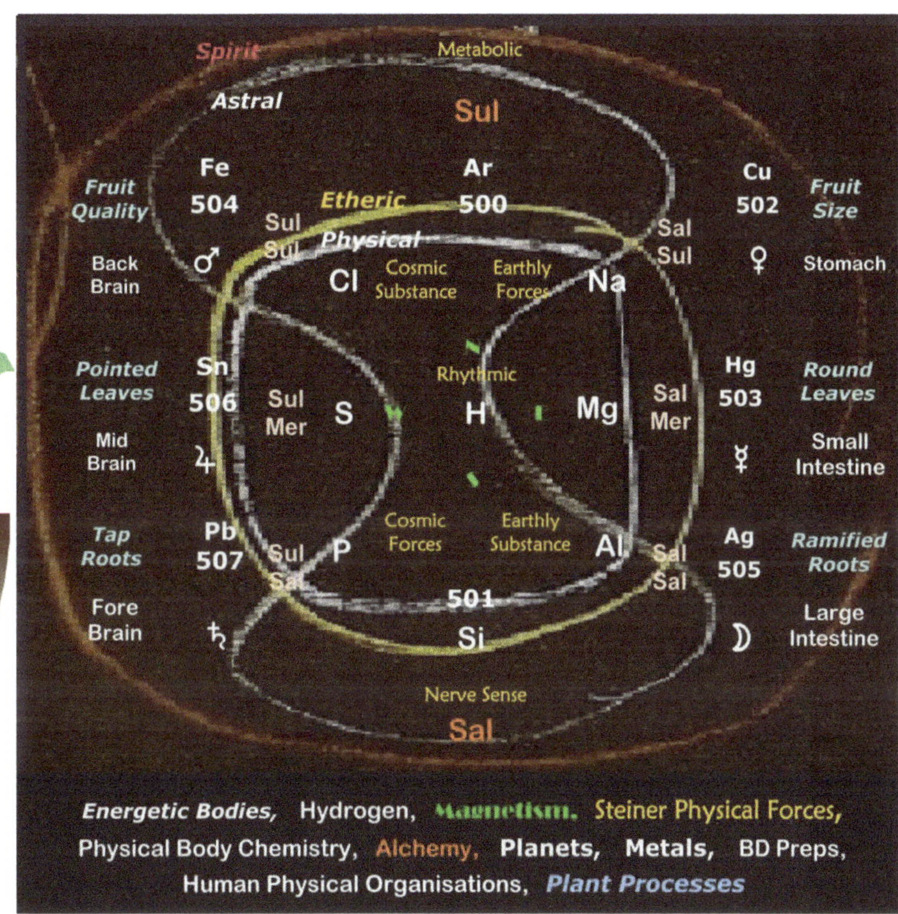

SPIRIT Warmth Cosmic Forces Indirect Outer Pl. Clay	Cambium Salamanders ↓	Saturn	P	Valerian 507	Strengthens the Spirit
	Sylphs	Jupiter	S	Dandelion 506	Merge Spirit and Astral towards the Physical
ASTRAL Light Cosmic Substance Direct Outer Pl. Sand		Mars	Cl	Nettle 504	Harmonise the Astrality
	Undines Life Sap ↓	Sun	Ar	Equisetum 508	Astral stimulates the Etheric within the metabolism
ETHERIC Chemical Earthly Forces Direct Inner Pl. Humus		Venus	Na	Yarrow 502	Opens the Etheric to the Astral
	Gnomes Wood Sap	Mercury	Mg	Chamomile 503	Stimulates the Etheric
PHYSICAL Life Earthly Substance Indirect Inner Pl. Cations	↑	Moon	Al	Oak Bark 505	Etheric binds to the Physical
	500	Earth	Si	Quartz 501	Spirit binds to the Physical

The Seasonal Complex 1

Magnetic North

The Rosetta Stone represents
the energies relationships

it moves
through the Seasons

The Seasons move
in an anti clockwise direction
when facing North

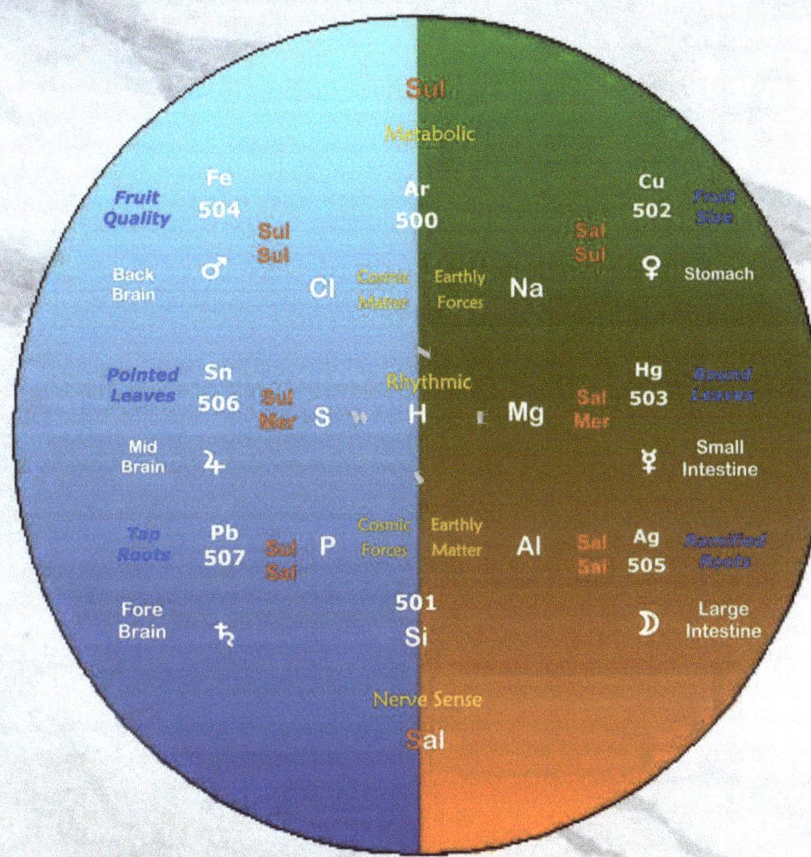

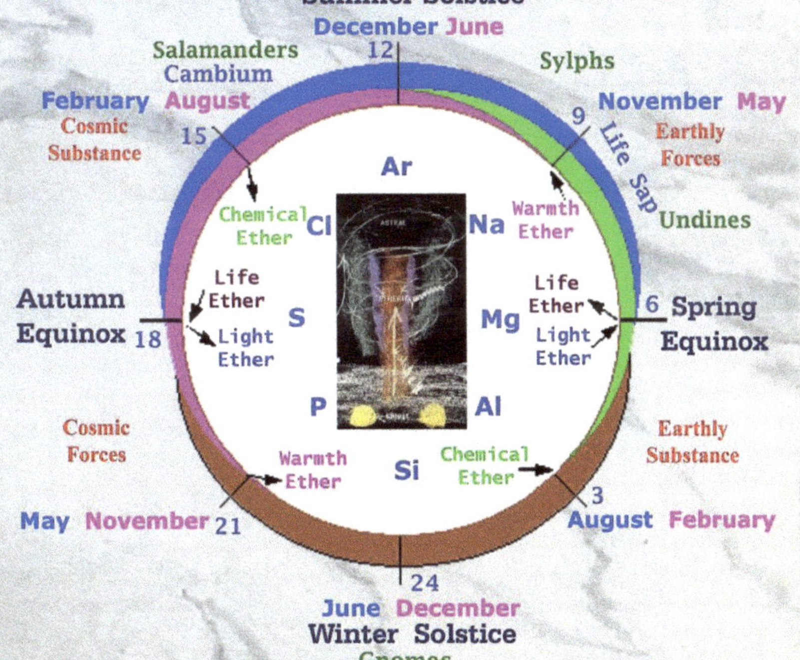

RS 'The Plant', G. Chemistry, Ethers, Time, Months, Seasons, Elementals , Cambium, Physical Formative Forces

The Seasonal Complex 2

Magnetic North

South Zodiacs

Follow the Seasons

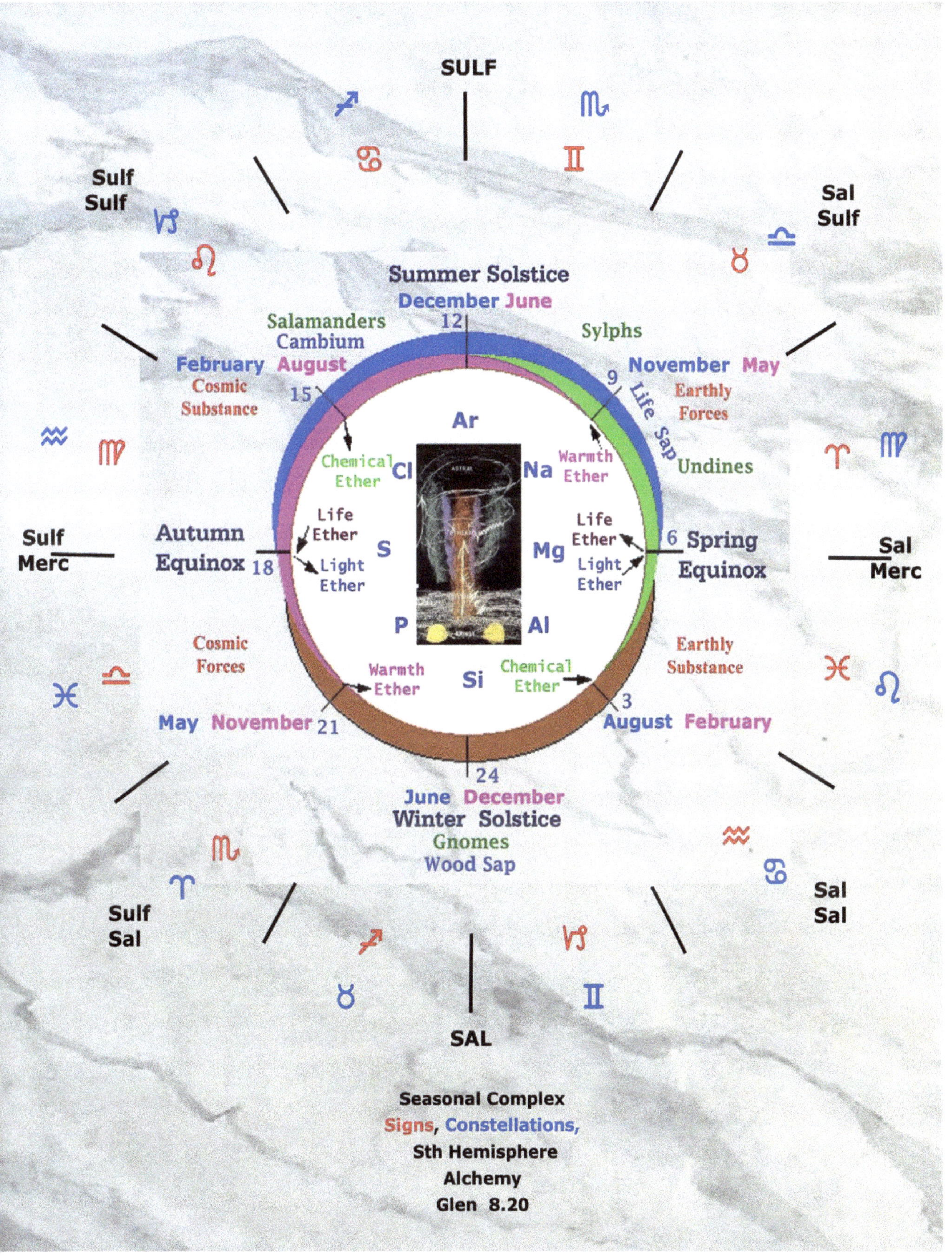

Seasonal Complex
Signs, Constellations,
Sth Hemisphere
Alchemy
Glen 8.20

Fungal Problems

Facing South

Pest Problems

Animal Nutrition

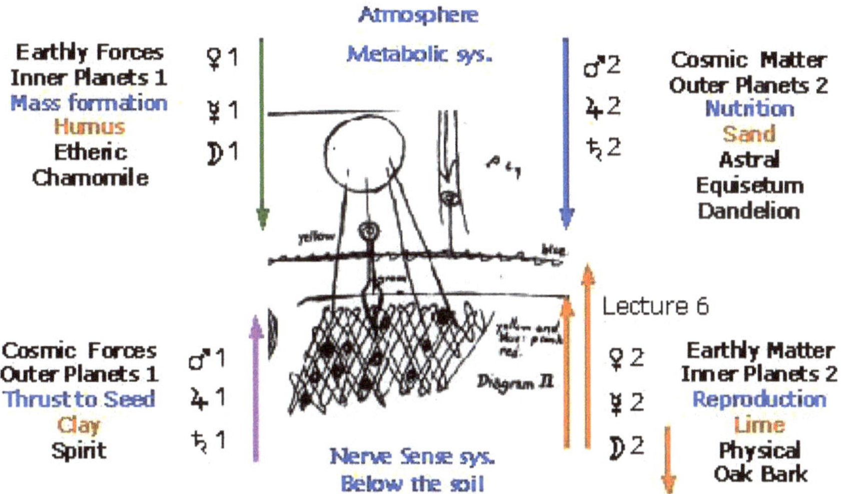

Atmosphere
Metabolic sys.
Earthly Forces
Inner Planets 1
Mass formation
Humus
Etheric
Chamomile
♀1
☿1
☽1
♂2
♃2
♄2
Cosmic Matter
Outer Planets 2
Nutrition
Sand
Astral
Equisetum
Dandelion
yellow
blue
Diagram II
Lecture 6
Cosmic Forces
Outer Planets 1
Thrust to Seed
Clay
Spirit
♂1
♃1
♄1
Nerve Sense sys.
Below the soil
♀2
☿2
☽2
Earthly Matter
Inner Planets 2
Reproduction
Lime
Physical
Oak Bark

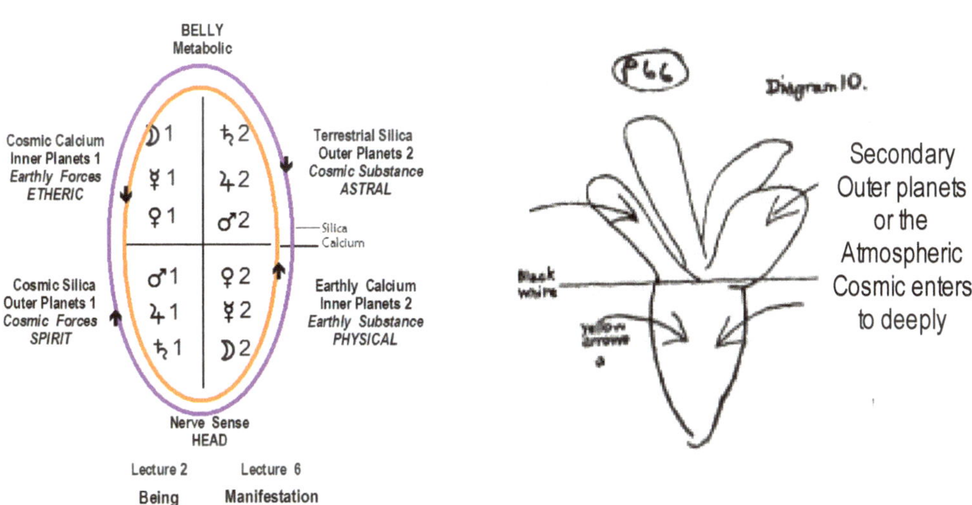

BELLY
Metabolic
Cosmic Calcium
Inner Planets 1
Earthly Forces
ETHERIC
☽1 ☿1 ♀1
♄2 ♃2 ♂2
Terrestrial Silica
Outer Planets 2
Cosmic Substance
ASTRAL
Silica
Calcium
Cosmic Silica
Outer Planets 1
Cosmic Forces
SPIRIT
♂1 ♃1 ♄1
♀2 ☿2 ☽2
Earthly Calcium
Inner Planets 2
Earthly Substance
PHYSICAL
Nerve Sense
HEAD
Lecture 2
Being
Lecture 6
Manifestation
P66
Diagram 10.
Secondary
Outer planets
or the
Atmospheric
Cosmic enters
to deeply
black
white
yellow
arrows

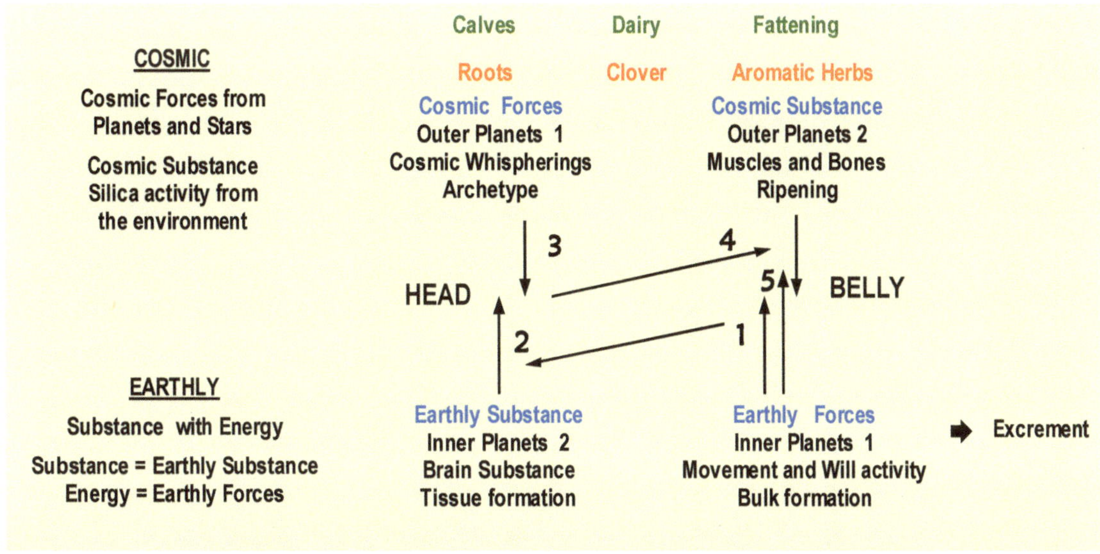

COSMIC
Cosmic Forces from Planets and Stars
Cosmic Substance Silica activity from the environment
Calves
Dairy
Fattening
Roots
Clover
Aromatic Herbs
Cosmic Forces
Outer Planets 1
Cosmic Whispherings
Archetype
Cosmic Substance
Outer Planets 2
Muscles and Bones
Ripening
3
4
HEAD
5
BELLY
2
1
EARTHLY
Substance with Energy
Substance = Earthly Substance
Energy = Earthly Forces
Earthly Substance
Inner Planets 2
Brain Substance
Tissue formation
Earthly Forces
Inner Planets 1
Movement and Will activity
Bulk formation
Excrement

Biodynamic Chemistry 2

Facing South

Given this is a story of universal order

Chemistry, must be
a manifestation of this order.

The three planes of chemical activity

The three groups of elements
Majors, Transitions, Radioactive Elements
as Three Layers

The Manifest layer

The Chartres Labyrinth

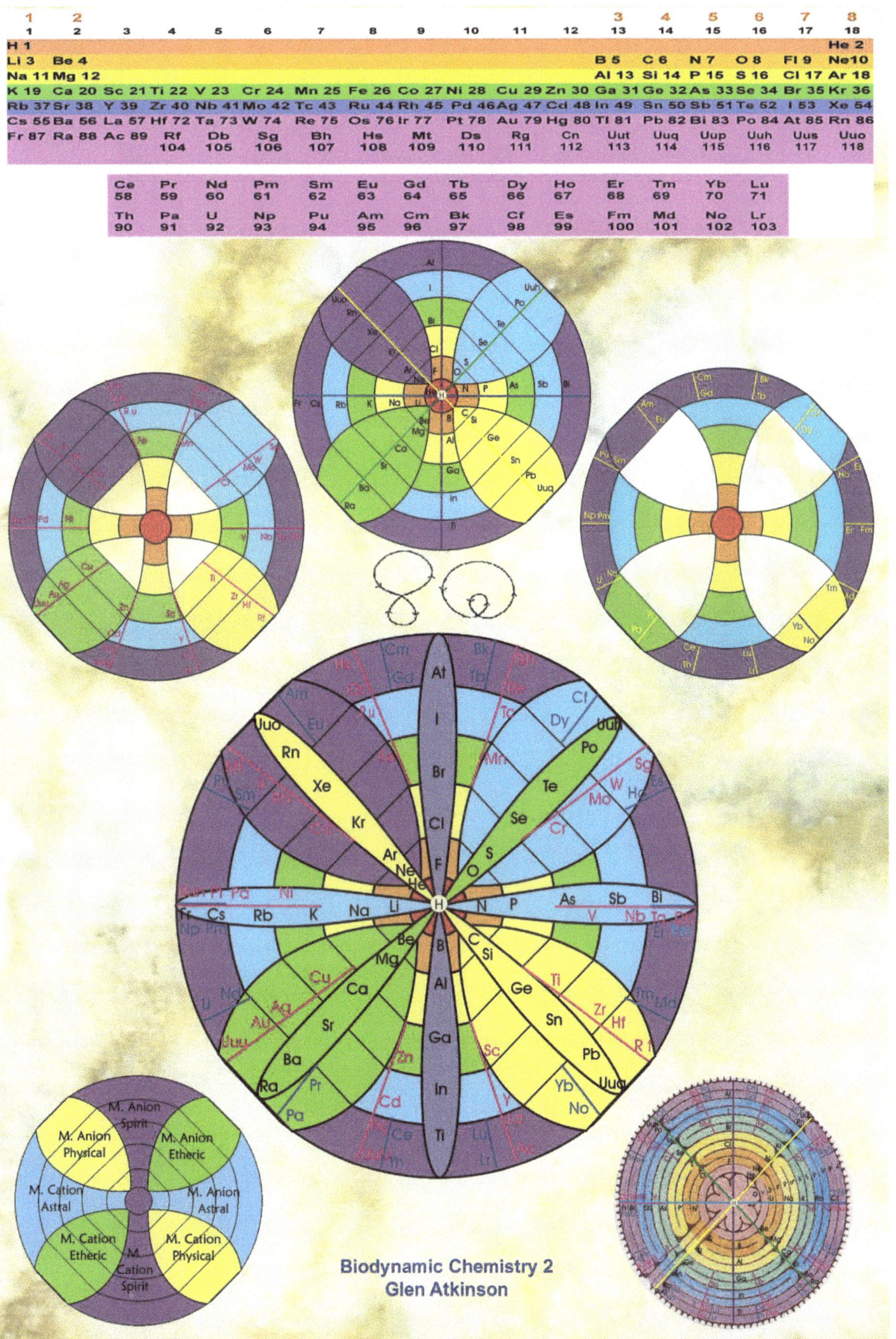

1 2 3 4 5 6 7 8
1 2 3 4 5 6 7 8 9 10 11 12 13 14 15 16 17 18
H 1 He 2
Li 3 Be 4 B 5 C 6 N 7 O 8 Fl 9 Ne10
Na 11 Mg 12 Al 13 Si 14 P 15 S 16 Cl 17 Ar 18
K 19 Ca 20 Sc 21 Ti 22 V 23 Cr 24 Mn 25 Fe 26 Co 27 Ni 28 Cu 29 Zn 30 Ga 31 Ge 32 As 33 Se 34 Br 35 Kr 36
Rb 37 Sr 38 Y 39 Zr 40 Nb 41 Mo 42 Tc 43 Ru 44 Rh 45 Pd 46 Ag 47 Cd 48 In 49 Sn 50 Sb 51 Te 52 I 53 Xe 54
Cs 55 Ba 56 La 57 Hf 72 Ta 73 W 74 Re 75 Os 76 Ir 77 Pt 78 Au 79 Hg 80 Tl 81 Pb 82 Bi 83 Po 84 At 85 Rn 86
Fr 87 Ra 88 Ac 89 Rf 104 Db 105 Sg 106 Bh 107 Hs 108 Mt 109 Ds 110 Rg 111 Cn 112 Uut 113 Uuq 114 Uup 115 Uuh 116 Uus 117 Uuo 118
Ce 58 Pr 59 Nd 60 Pm 61 Sm 62 Eu 63 Gd 64 Tb 65 Dy 66 Ho 67 Er 68 Tm 69 Yb 70 Lu 71
Th 90 Pa 91 U 92 Np 93 Pu 94 Am 95 Cm 96 Bk 97 Cf 98 Es 99 Fm 100 Md 101 No 102 Lr 103
M. Anion Spirit
M. Anion Physical
M. Anion Etheric
M. Cation Astral
M. Anion Astral
M. Cation Etheric
M. Cation Physical
M. Cation Spirit
Biodynamic Chemistry 2
Glen Atkinson

Biodynamic Chemistry

Facing South

The chemical elements activity must be
described by the energetic glossary

A further step
similar to the ag course lemniscate
is taken when we see
RS story of chemistry based upon H,N,O,C

He describes a chemistry where
the elements work into
the physical level of manifestation,
through their roles as
polarised charged particles.
see the bottom left diagram

Look for the Three Worlds of activity
for each of the elements.

The Gyroscopic Periodic Table

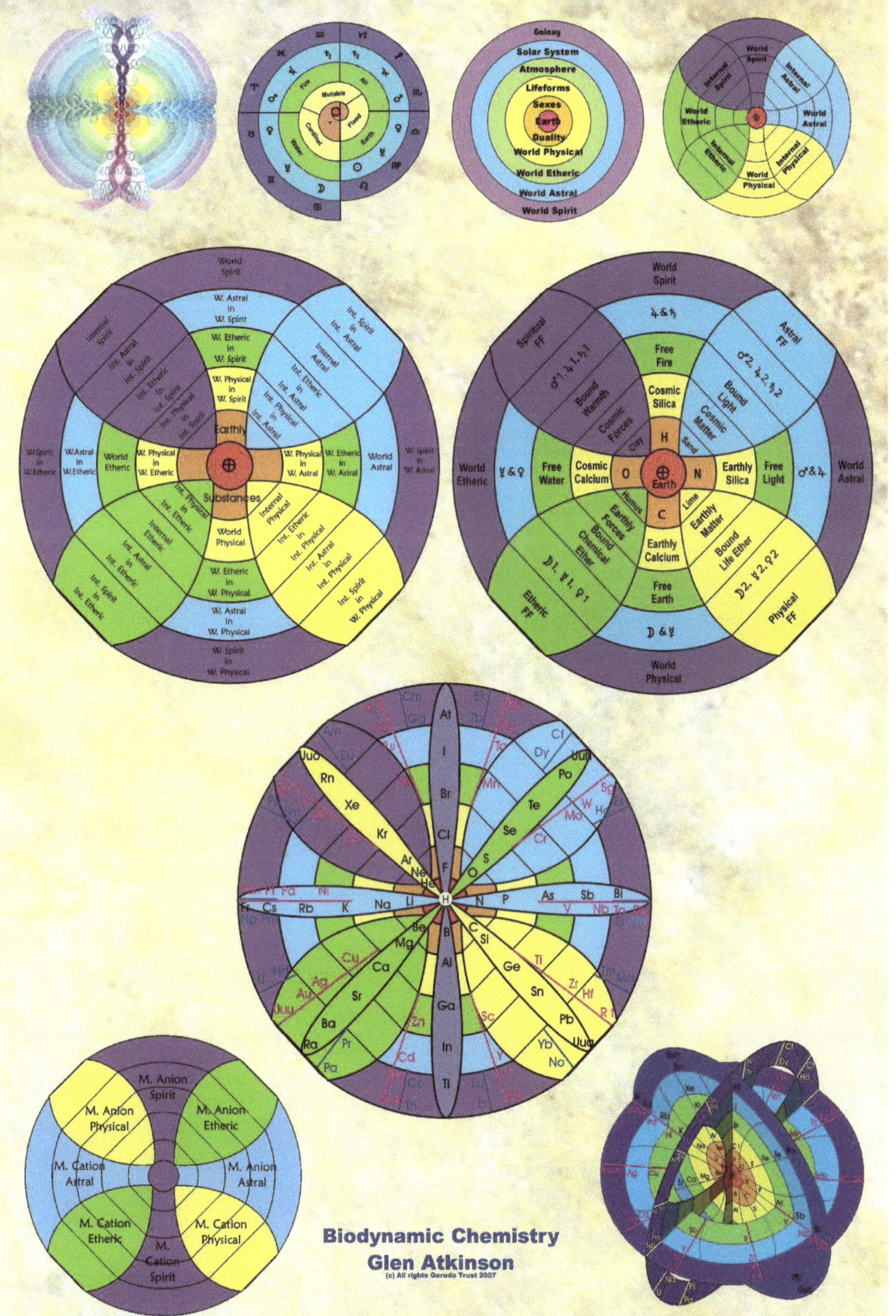
Galaxy
Solar System
Atmosphere
Lifeforms
Sexes
Earth
Duality
World Physical
World Etheric
World Astral
World Spirit
Mutable
Fixed
Cardinal
Fire
Air
Earth
Water
World Spirit
Internal Spirit
Internal Astral
World Astral
World Etheric
Internal Etheric
World Physical
Internal Physical
World Spirit
Internal Spirit
W. Astral in W. Spirit
W. Etheric in W. Spirit
W. Physical in W. Spirit
Int. Spirit in Int. Astral
Internal Astral
Int. Astral in Int. Spirit
Int. Etheric in Int. Spirit
Int. Physical in Int. Spirit
Int. Etheric in Int. Astral
Int. Physical in Int. Astral
Earthly
Substances
W.Spirit in W.Etheric
W.Astral in W.Etheric
World Etheric
W. Physical in W. Etheric
W. Physical in W. Astral
W. Etheric in W. Astral
World Astral
W. Spirit in W. Astral
Int. Physical in Int. Etheric
Internal Etheric
Int. Astral in Int. Etheric
Int. Spirit in Int. Etheric
Internal Physical
Int. Etheric in Int. Physical
Int. Astral in Int. Physical
Int. Spirit in W. Physical
World Physical
W. Etheric in W. Physical
W. Astral in W. Physical
W. Spirit in W. Physical
World Spirit
Spiritual FF
Astral FF
Free Fire
Cosmic Silica
Bound Warmth
Cosmic Forces
Clay
Bound Light
Cosmic Matter
Sand
H
World Etheric
Free Water
Cosmic Calcium
O
Earth
N
Earthly Silica
Free Light
World Astral
Humus
Lime
C
Earthly Forces
Bound Chemical Ether
Earthly Matter
Bound Life Ether
Earthly Calcium
Free Earth
Etheric FF
Physical FF
World Physical
M. Anion Spirit
M. Anion Physical
M. Anion Etheric
M. Cation Astral
M. Anion Astral
M. Cation Etheric
M. Cation Physical
M. Cation Spirit
Biodynamic Chemistry
Glen Atkinson
(c) All rights Garuda Trust 2007

All In One

Facing South

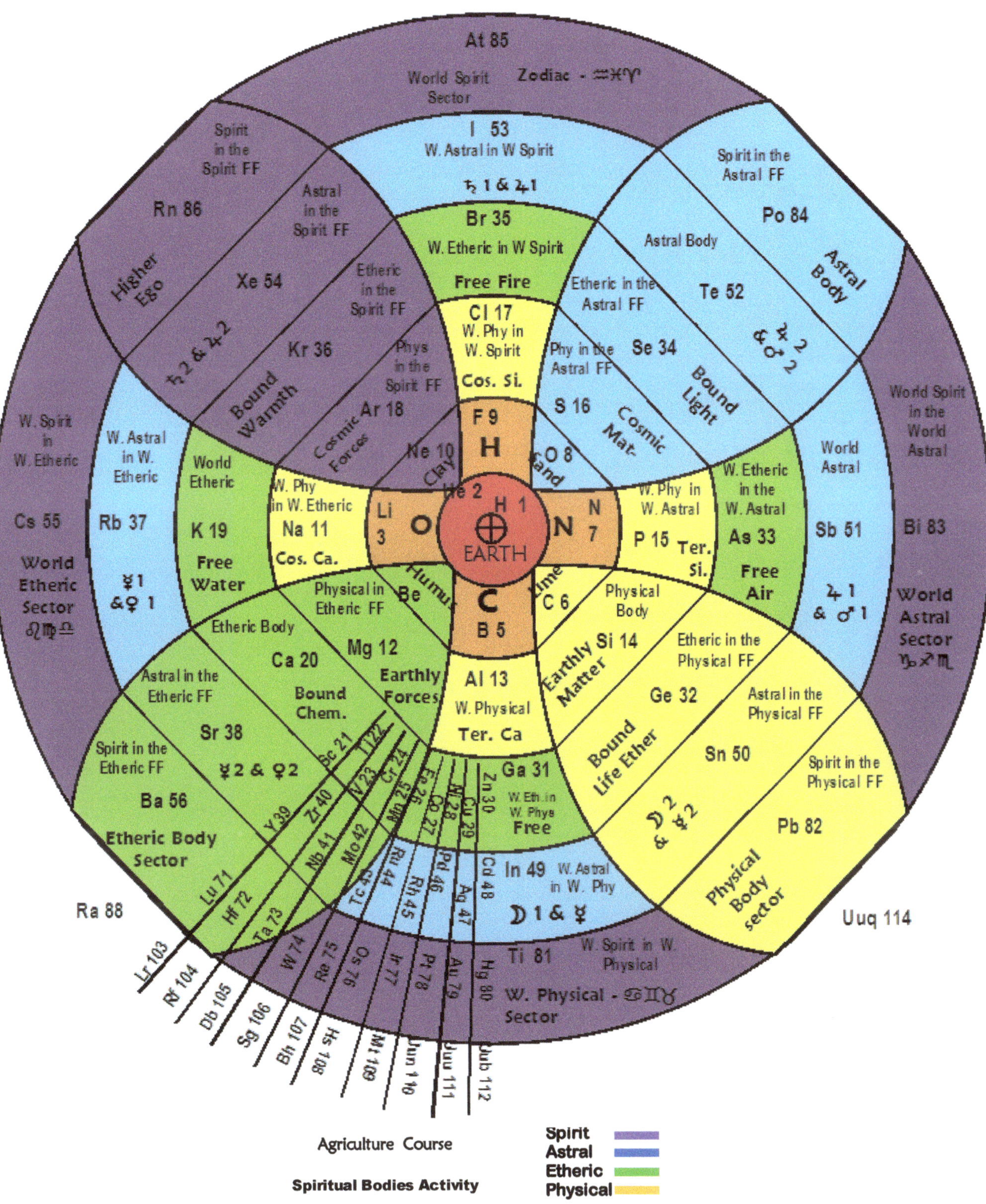
At 85
World Spirit Sector
Zodiac - ♒♓♈
I 53
W. Astral in W Spirit
♄ 1 & ♃ 1
Br 35
W. Etheric in W Spirit
Free Fire
Cl 17
W. Phy in W. Spirit
Cos. Si.
F 9
H
O 8
Sand
Spirit in the Spirit FF
Rn 86
Higher Ego
Astral in the Spirit FF
Xe 54
♄ 2 & ♃ 2
Etheric in the Spirit FF
Kr 36
Bound Warmth
Phys in the Spirit FF
Ar 18
Cosmic Forces
Ne 10
Clay
Spirit in the Astral FF
Po 84
Astral Body
Astral Body
Te 52
♃ 2 & ♂ 2
Etheric in the Astral FF
Se 34
Bound Light
Phy in the Astral FF
S 16
Cosmic Mat.
W. Spirit in W. Etheric
Cs 55
World Etheric Sector
♌♍♎
W. Astral in W. Etheric
Rb 37
☿ 1 & ♀ 1
World Etheric
K 19
Free Water
W. Phy in W. Etheric
Na 11
Cos. Ca.
Li 3
O
He 2
H 1
EARTH
N
N 7
C
B 5
World Spirit in the World Astral
Bi 83
World Astral Sector
♑♐♏
World Astral
Sb 51
♃ 1 & ♂ 1
W. Etheric in the W. Astral
As 33
Free Air
W. Phy in W. Astral
P 15
Ter. Si.
Physical in Etheric FF
Be
Humus
Lime
C 6
Physical Body
Etheric Body
Ca 20
Bound Chem.
Mg 12
Earthly Forces
Al 13
W. Physical
Ter. Ca
Astral in the Etheric FF
Sr 38
☿ 2 & ♀ 2
Spirit in the Etheric FF
Ba 56
Etheric Body Sector
Earthly Matter
Si 14
Etheric in the Physical FF
Ge 32
Bound Life Ether
Astral in the Physical FF
Sn 50
☽ 2 & ☿ 2
Spirit in the Physical FF
Pb 82
Physical Body sector
Sc 21
Ti 22
V 23
Cr 24
Mn 25
Fe 26
Co 27
Ni 28
Cu 29
Zn 30
Ga 31
W. Eth. in W. Phys
Free
Y 39
Zr 40
Nb 41
Mo 42
Tc 43
Ru 44
Rh 45
Pd 46
Ag 47
Cd 48
In 49
W. Astral in W. Phy
☽ 1 & ☿
Lu 71
Hf 72
Ta 73
W 74
Re 75
Os 76
Ir 77
Pt 78
Au 79
Hg 80
Tl 81
W. Spirit in W. Physical
W. Physical - ♋♊♉
Sector
Ra 88
Uuq 114
Lr 103
Rf 104
Db 105
Sg 106
Bh 107
Hs 108
Mt 109
Uun 110
Uuu 111
Uub 112
Agriculture Course
Spiritual Bodies Activity
Spirit
Astral
Etheric
Physical

Glenological Chemistry

Facing South

Three Groups of Chemical Elements

Working on

Three Layers of Creation

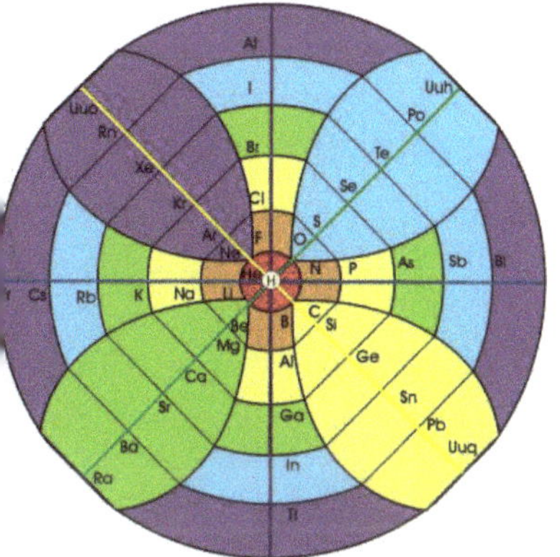

Prime Vertical

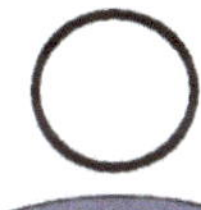

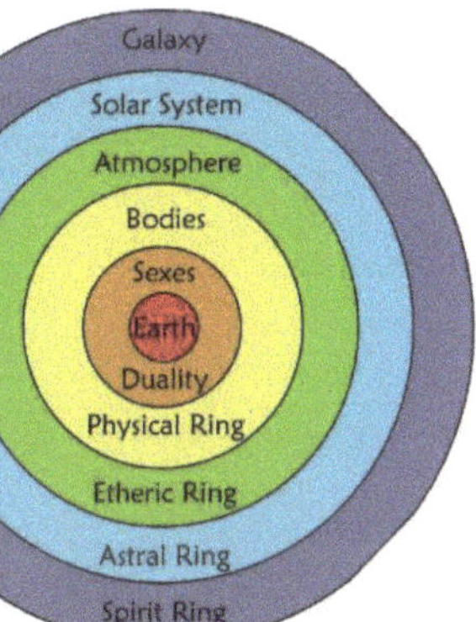

Stage 1
Archetype
Thesis
Passive
The Constant Field
Spirit
Saturn
Cosmic

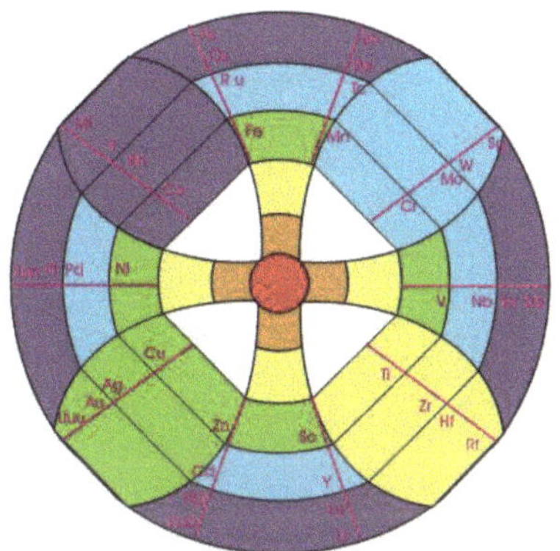

Horizontal

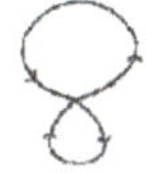

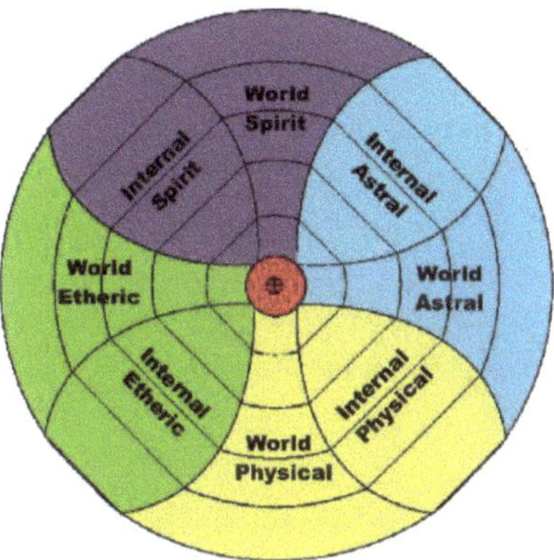

Stage 2
Forming
Antithesis
Pulsating
Surrounding the form
Soul
Jupiter
World and Internal

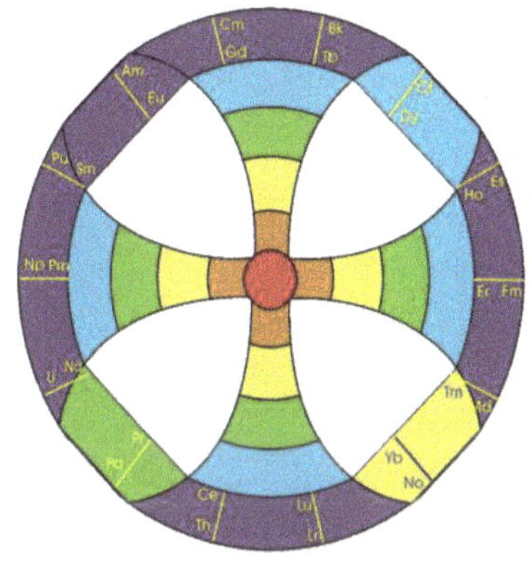

Second Vertical

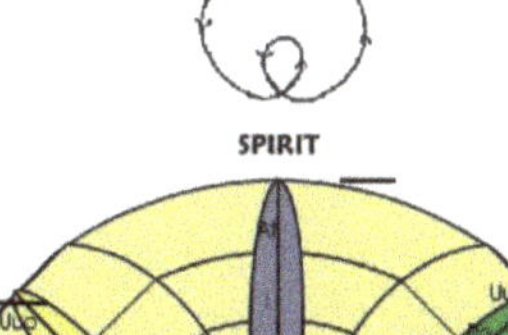

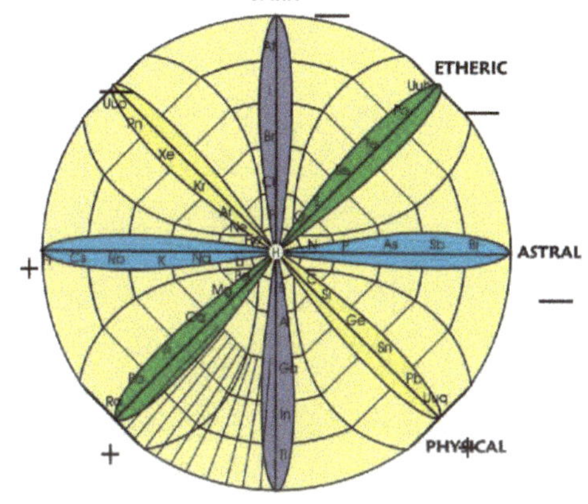

Stage 3
Manifestation
Synthesis
Enfolded
Form
Body
Chemistry

Transition Elements

Facing South

Look where the transition elements sit
on the energetic gyroscope

The transition elements
are most active as catalysts
of biological processes

Because we move closer to life
we take a lemniscate twist

The elemental order is read
K, Ca, Zn, Cu

Observe what happens to the elements
in their sequence of
hardness and melting points.

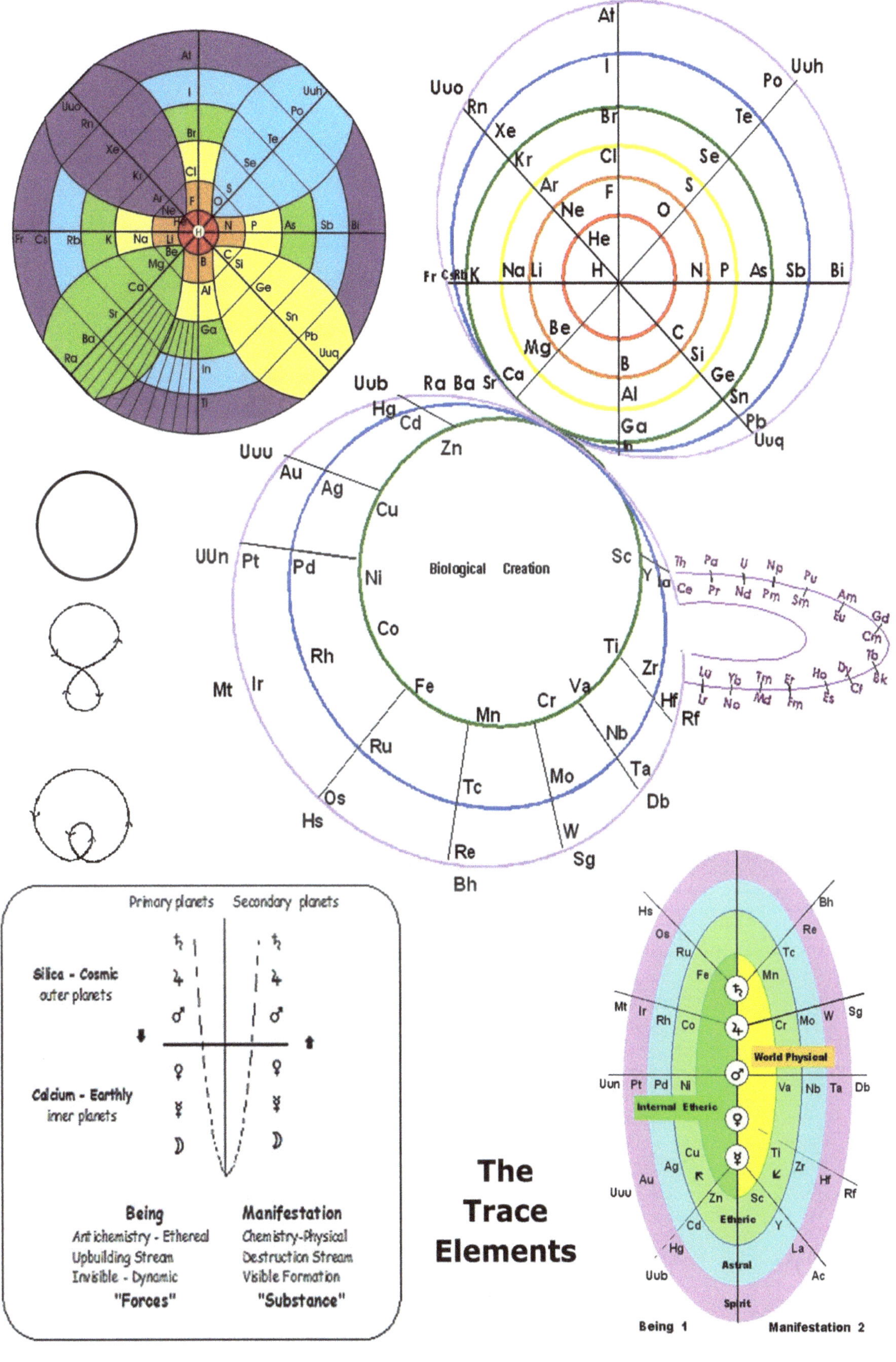

Biological Creation
Primary planets
Secondary planets
Silica - Cosmic
outer planets
Calcium - Earthly
inner planets
Being
Antichemistry - Ethereal
Upbuilding Stream
Invisible - Dynamic
"Forces"
Manifestation
Chemistry-Physical
Destruction Stream
Visible Formation
"Substance"
The
Trace
Elements
World Physical
Internal Etheric
Etheric
Astral
Spirit
Being 1
Manifestation 2

The Lanthanides

& Actinides

Magnetic North

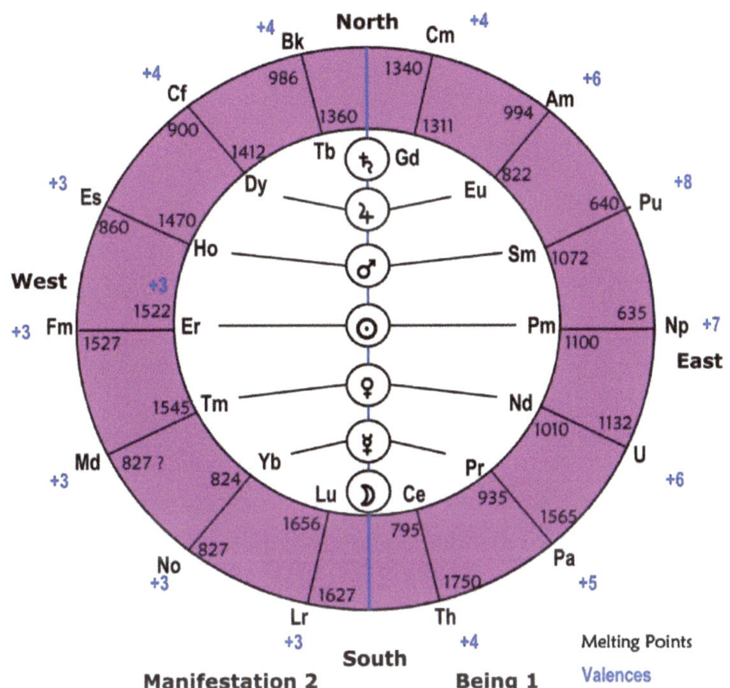

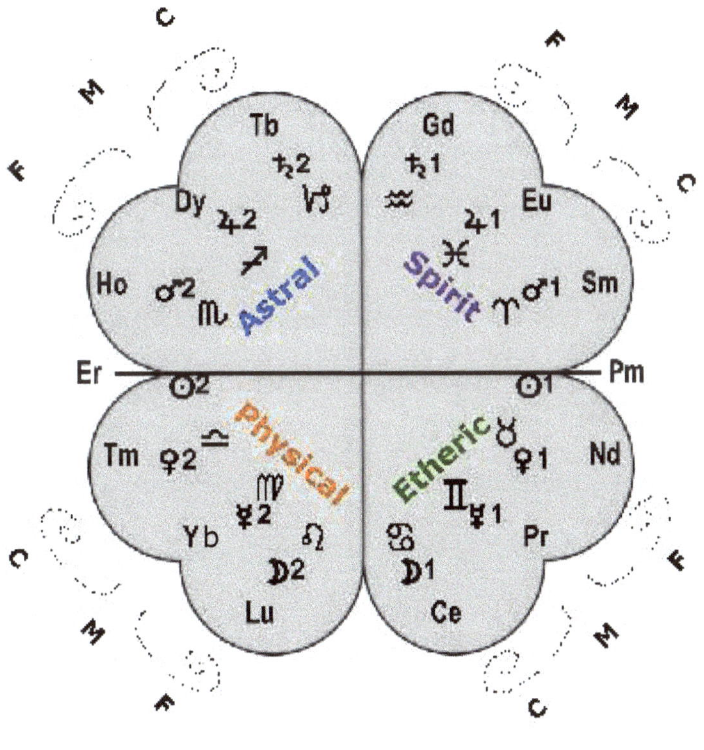

Planets
Constellations
Activities

C Metabolic
M Rhythmic
F Nerve Sense

3d Rings Periodic Table

Facing South

The three groups of chemical elements
are each on one axis plane
of the 3d gyroscope

The rings show the Cosmic activities

3d Arms Periodic Table

Stage two 3d activities

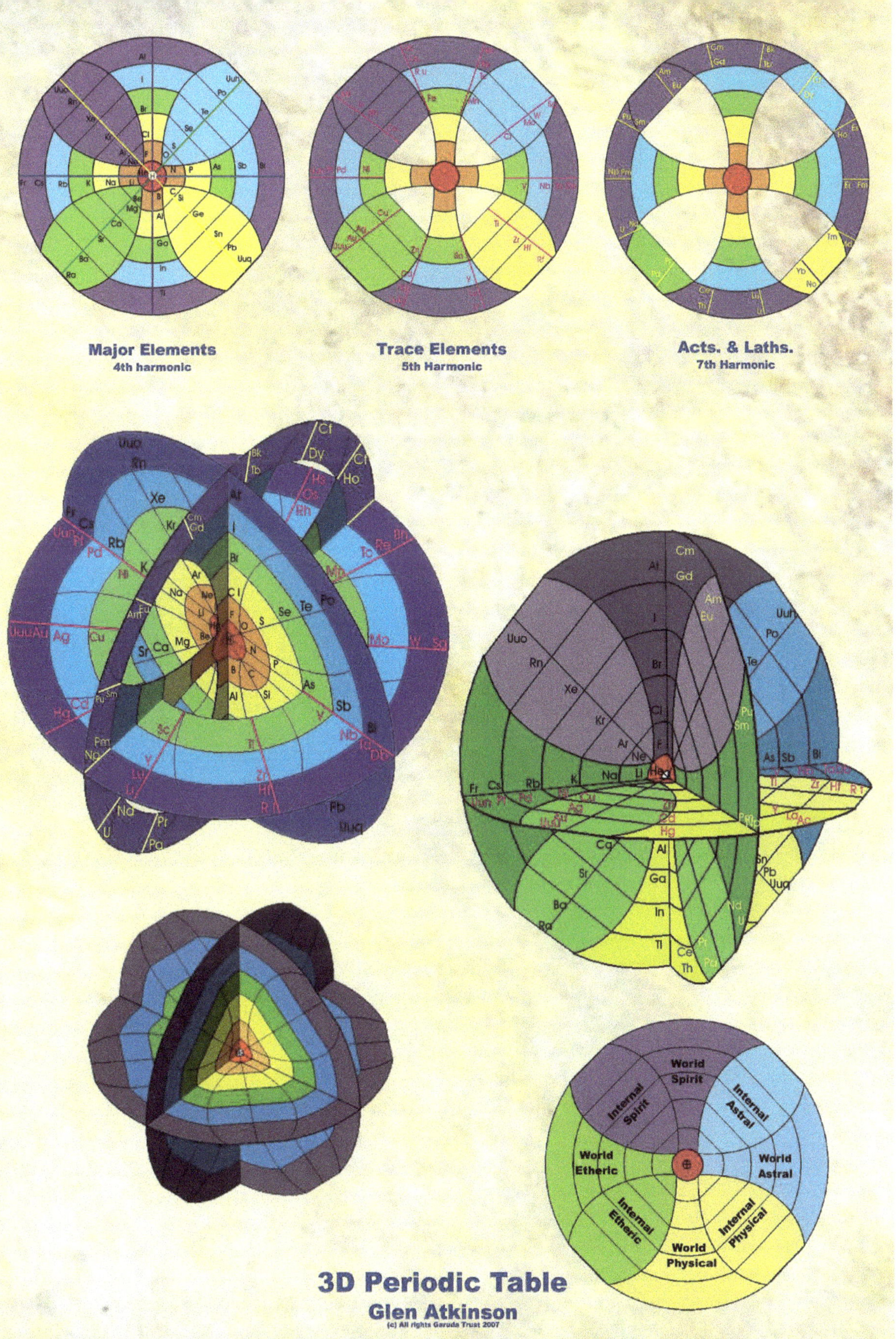
Major Elements
4th harmonic
Trace Elements
5th Harmonic
Acts. & Laths.
7th Harmonic
World Spirit
Internal Spirit
Internal Astral
World Etheric
World Astral
Internal Etheric
Internal Physical
World Physical
3D Periodic Table
Glen Atkinson
(c) All rights Garuda Trust 2007

Glenological Chemistry

Magnetic North

Look

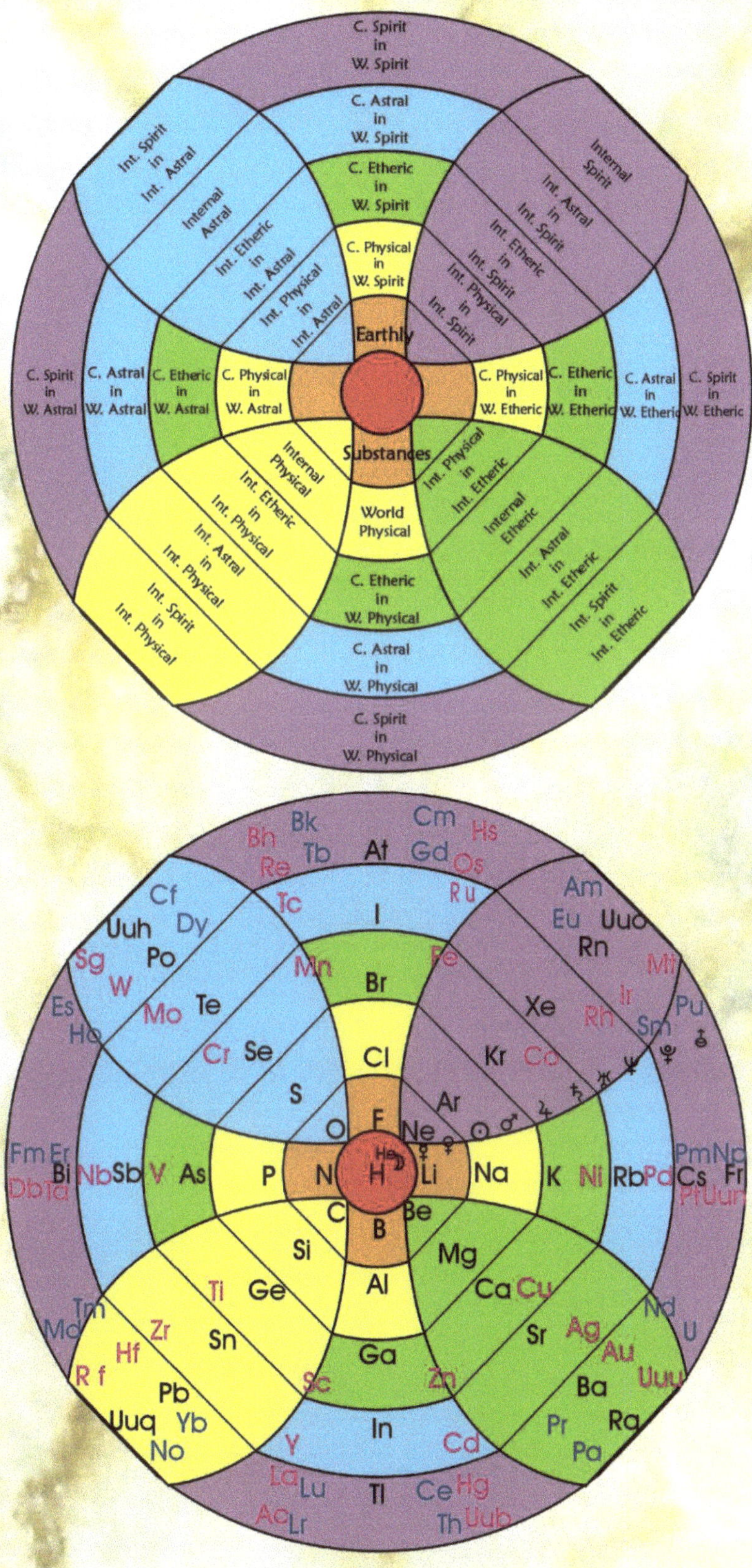

Glenological Chemistry North

Planets and Elements

Magnetic North

Lievegoed and the constellations
can be cross referenced
onto the gyroscope.

Observe how the elements are manifesting
the planetary and constellational activities.

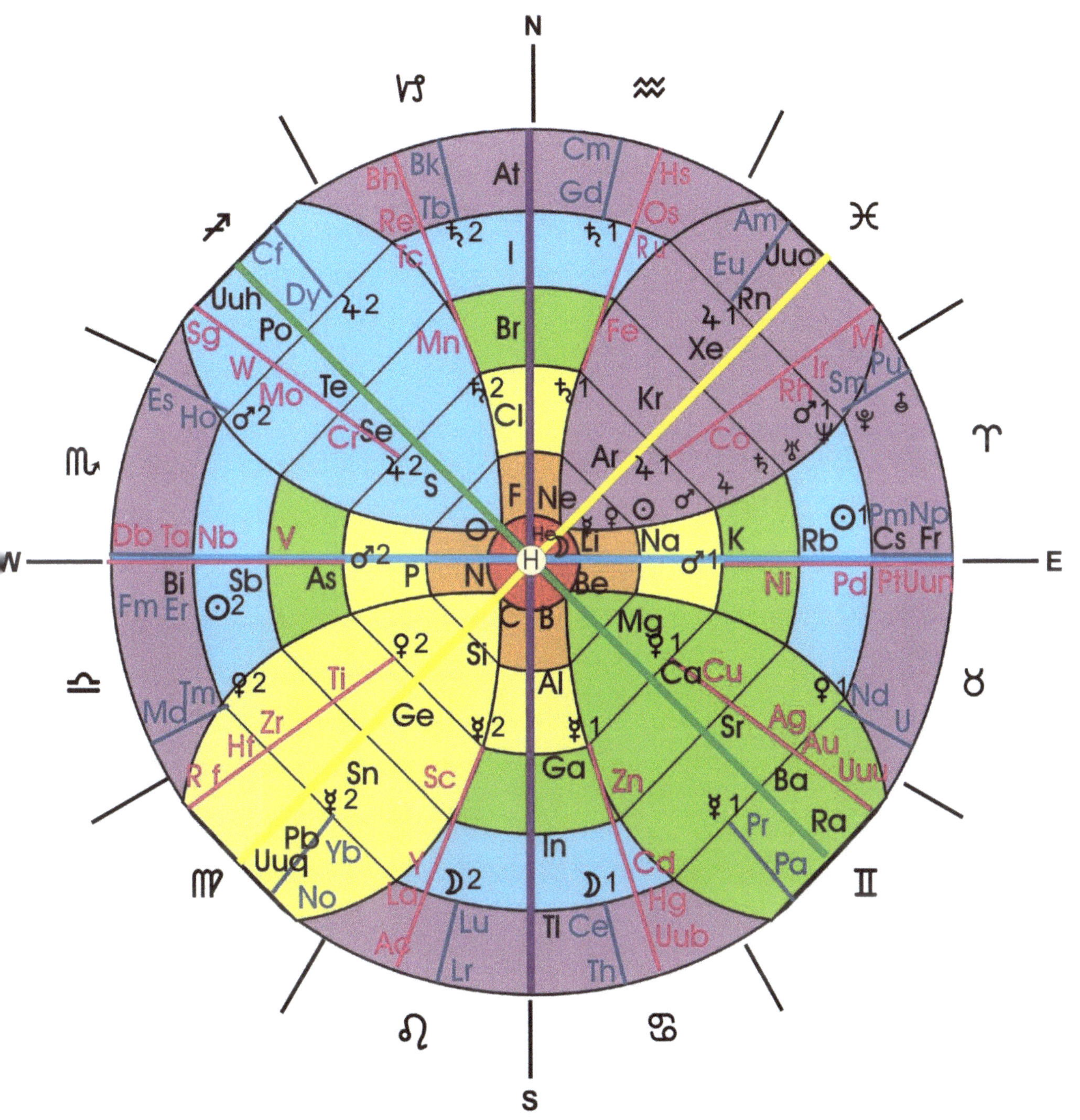

Glenological Chemistry,
Planets and Signs

Magnetic North Orientation

Metallic states

Magnetic North

If we are to look in greater depth into the energetic
organisation of manifest bodies
we need to orientate the gyroscope
upon the internal physical arm,
rather than the world physical arm

This places Carbon C
as the base of
physical manifestation

The metallic states show a
polarisation of the elements

3 Fold Plant

within the Periodic Tables organisation

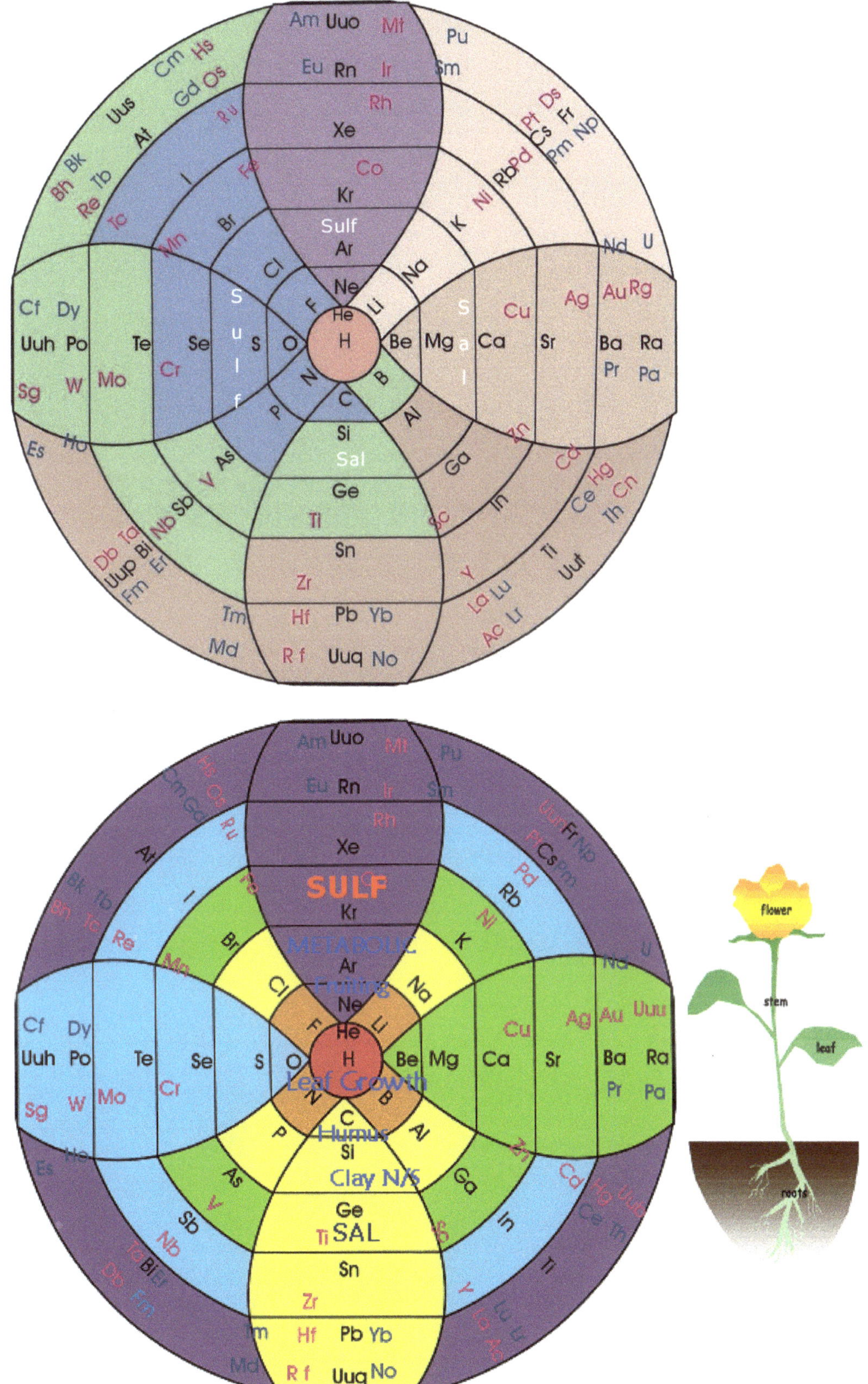
Sulf
S u l f
S a l
Sal
SULF
METABOLIC
Fruiting
Leaf Growth
Humus
Clay N/S
SAL
flower
stem
leaf
roots

Glenological Rosetta Stone 2

Magnetic North

The order of the elements that arises
from the internal physical body orientation
conforms to Dr Steiners stories
within the medical and agricultural lectures

Observe what this means
for how the elements work.

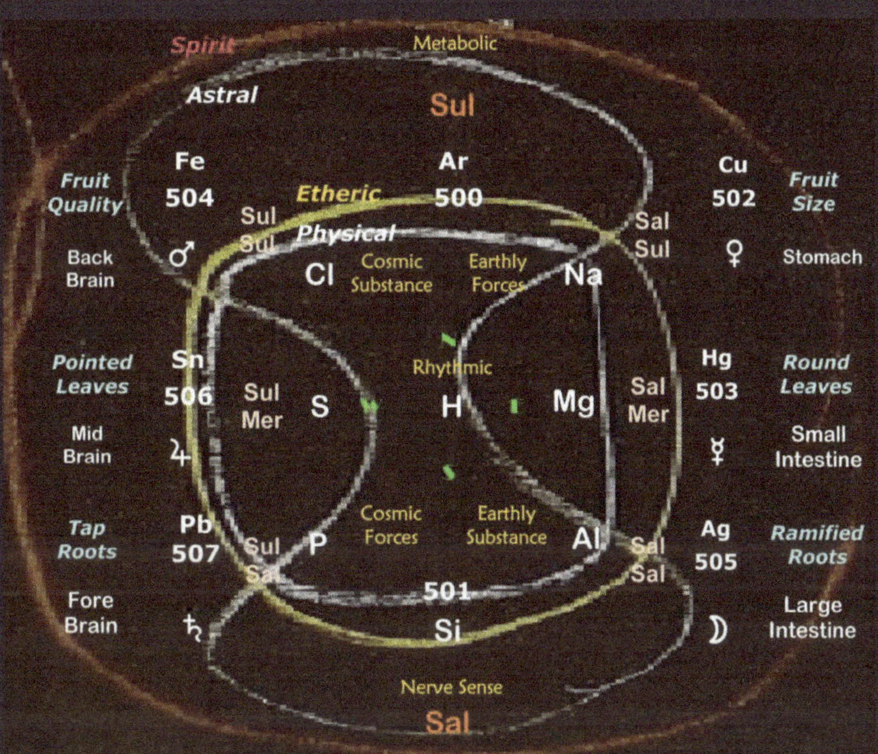

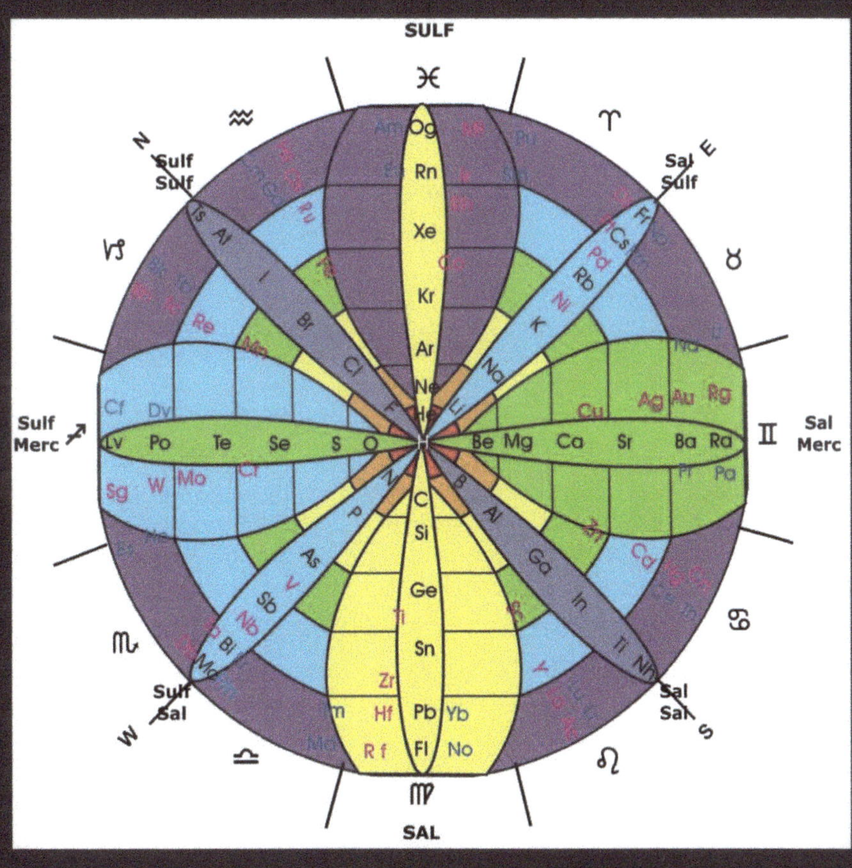

Glenological 'Rosetta Stone'

The Most Manifest

Facing South

Magnetic North

Gyroscopic Periodic Table

focused upon the

Internal Physical Arm

on the background of

Manifest Layer Organisation

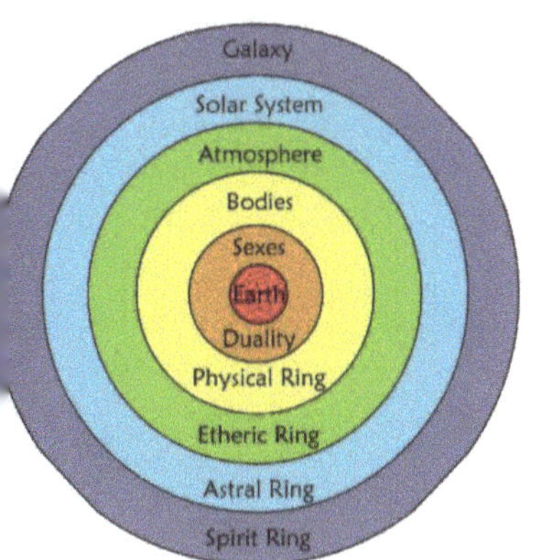

Stage 1
Archetype
Thesis
Passive
The Constant Field
Spirit
Saturn
Cosmic

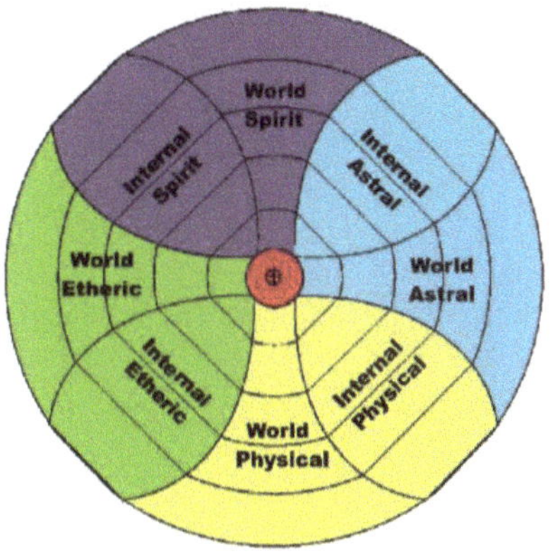

Stage 2
Forming
Antithesis
Pulsating
Surrounding the form
Soul
Jupiter
World and Internal

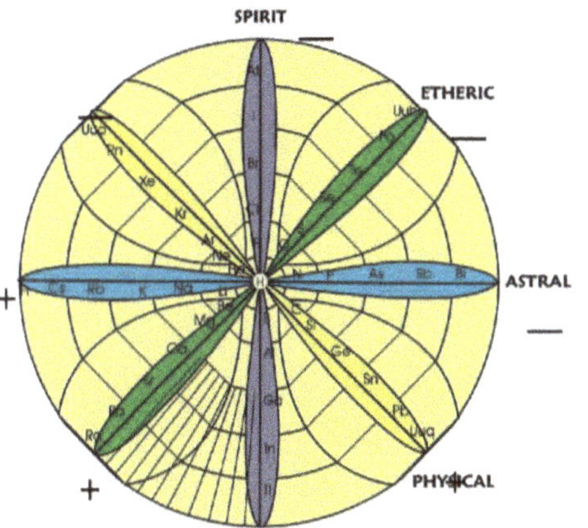

Stage 3
Manifestation
Synthesis
Enfolded
Form
Body
Chemistry

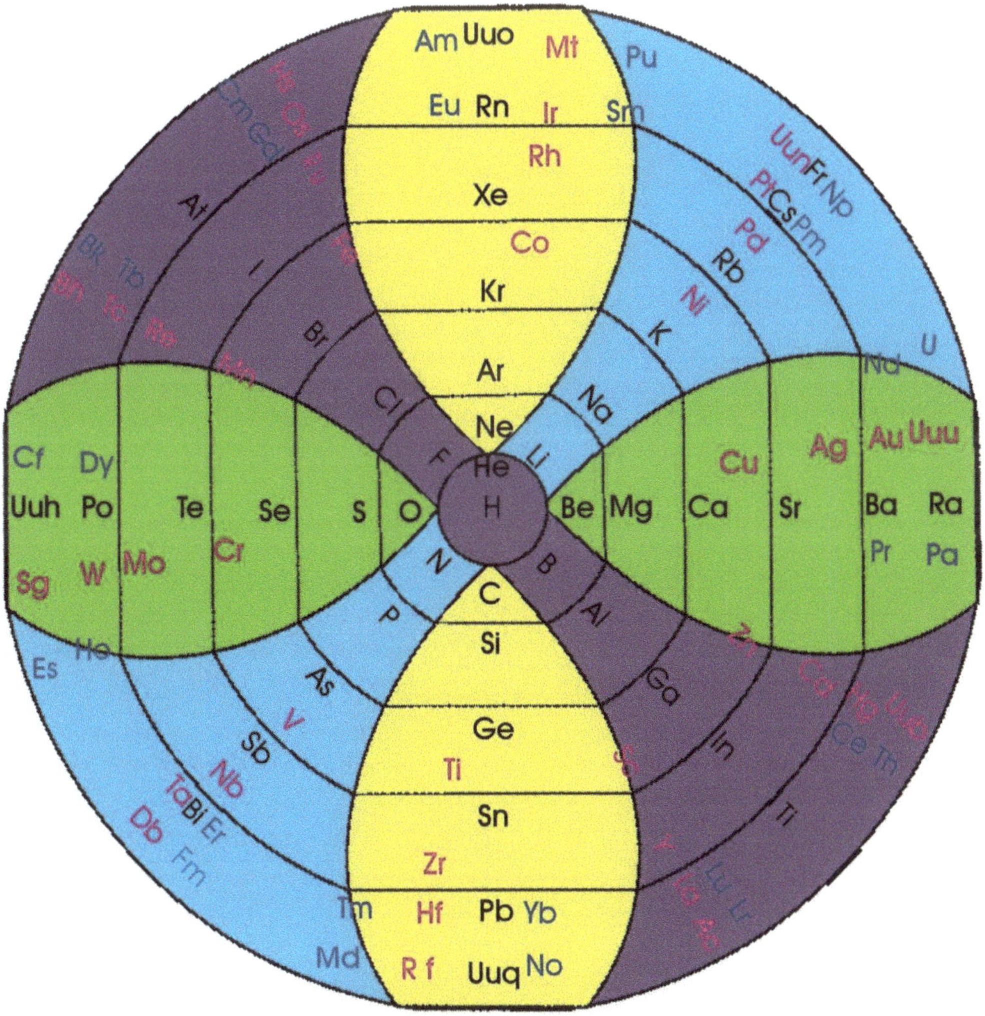

Chartres Labyrinth

Magnetic North

Chartres is aligned along
the NE SW axis ,
not the usual E W axis
of most cathedrals

So the labyrinth
has a SW opening

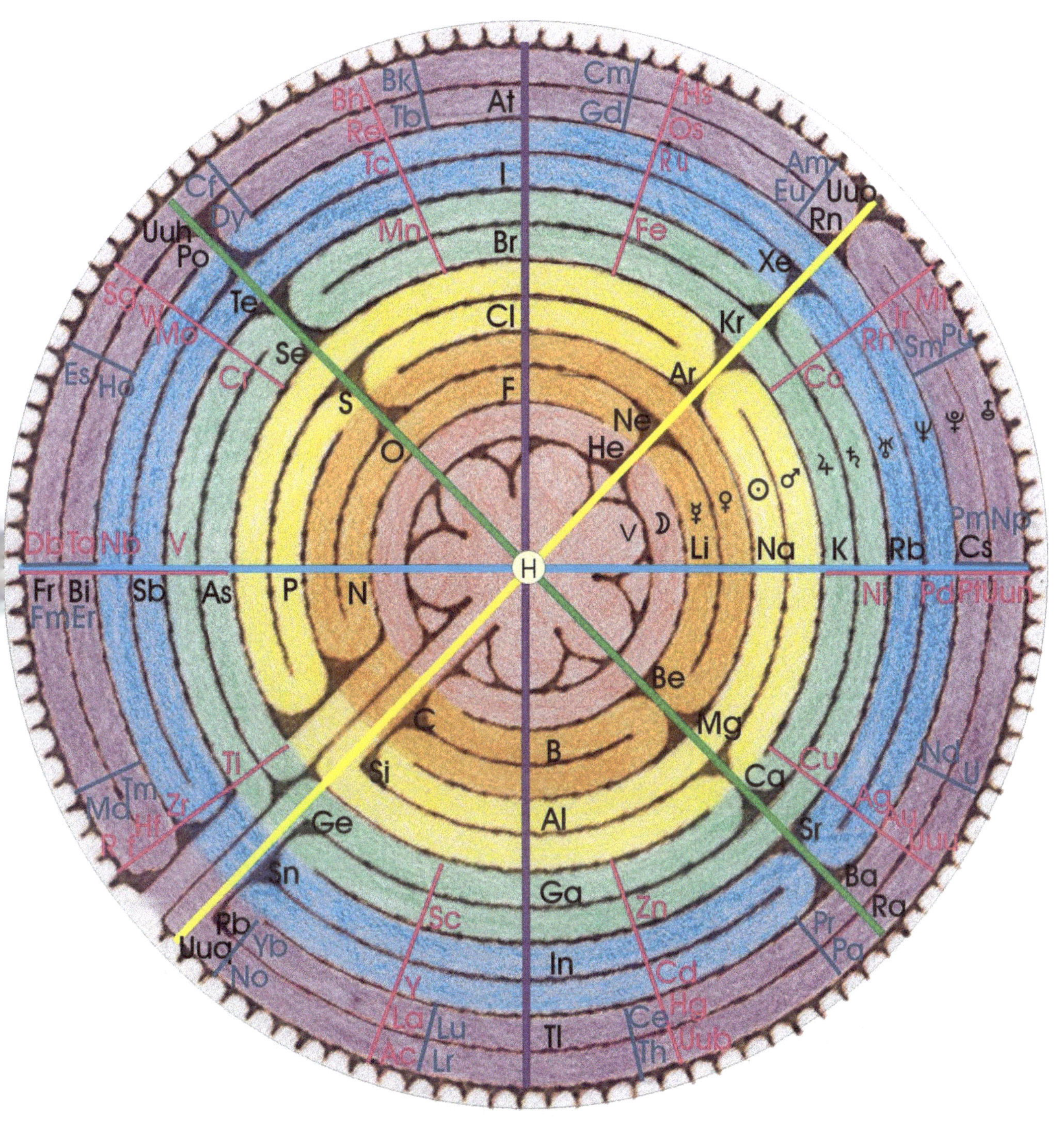

Chartres Labryinth – Periodic Table

Wells Cathedral

Chapter House

Magnetic North

Built by
the Knights Templar
in 1300 AD

who were Tried and
Sentenced to Death
in this building
in 1313 AD

This artwork
is dedicated to the message
they wish to convey
and encourage us
to use.

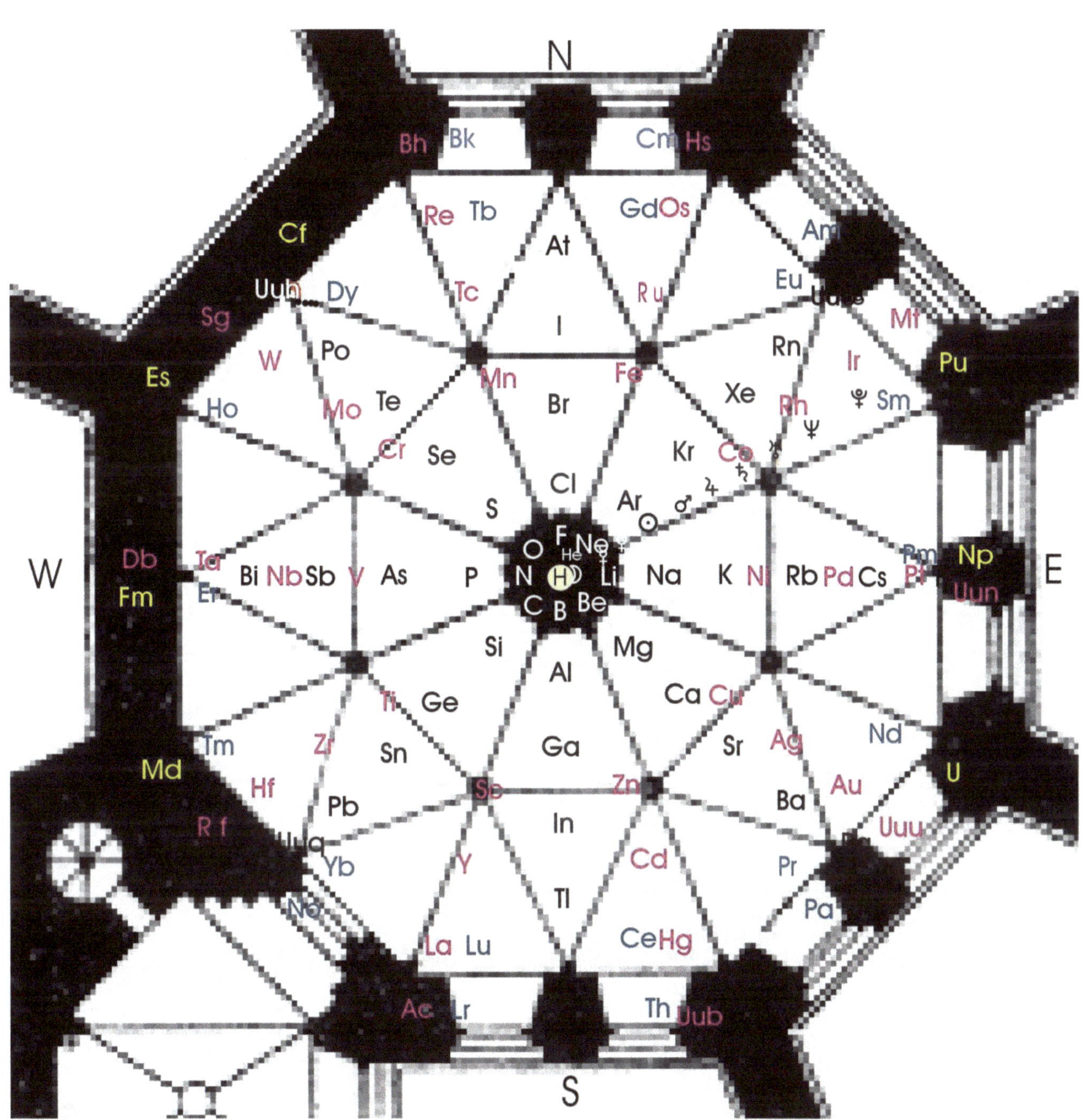
N
Bh Bk
Hs
Re Tb
GdOs
Cf
At
Am
Dy
Tc
Eu
Sg
I
Mt
W
Po
Mn
Fe
Rn
Ir
Pu
Es
Te
Br
Xe
Sm
Ho
Mo
Rh
Se
Kr
Cr
Cl
S
Ar
F Ne
He
Db
O
Np
W
Bi Nb Sb
As
P
N
H
Li
Na
K
Rb Pd Cs
E
Fm
C B Be
Si
Mg
Al
Ge
Ca
Tm
Sn
Ga
Sr
Ag
Nd
Md
Hf
Zn
U
Pb
Ba
Au
In
Cd
Yb
Pr
Tl
Pa
CeHg
La Lu
Ac
Th
S

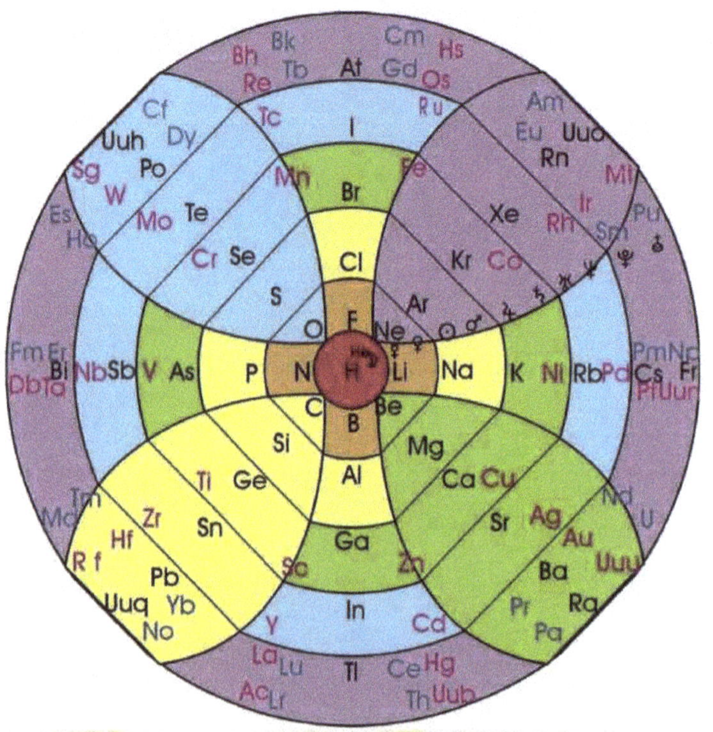

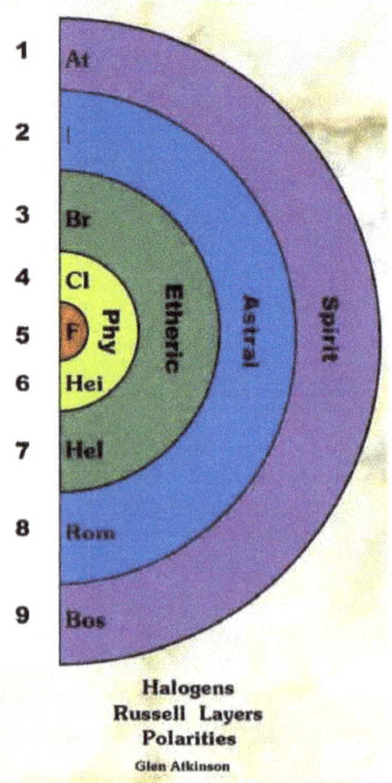

Halogens
Russell Layers
Polarities
Glen Atkinson

F
Hei
O Ne
Hel
Hal He
Buz Rom Gam
Del Bet
Bos
Blc Alp
N Lum Tra Bar Alb Ire Jam Mac Car Li
Tim Vij
Mar
Ath Ers
Qui Eye Pen
H Eth
Vin
C Be
Beb
B

Walter Russel Sub Atomic Particles
Circular North Layers Polarity
Glen Atkinson

There are 9 layers
Which may Polarise?

4 shells with a Centre
The Earthly Substances -
The Sphere of Polarity,
from which Life Emerges,
gets to play this Role.

Walter Russell
Sub Atomics
Circular
Layers Polarised
Circular Periodic

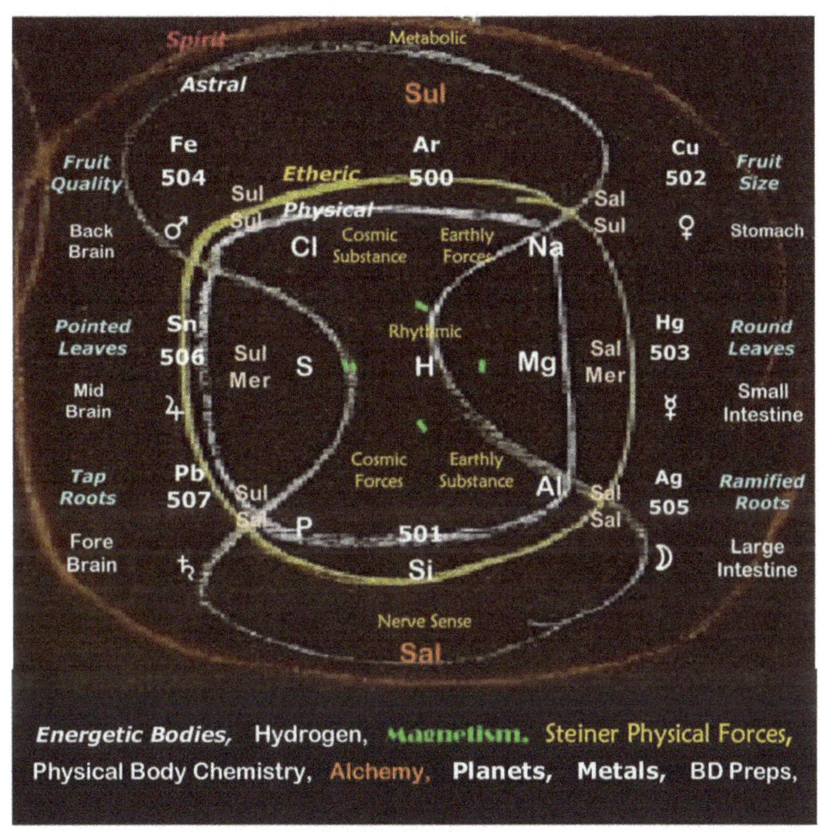
Spirit
Metabolic
Astral
Sul
Fe
504
Ar
500
Cu
502
Fruit Quality
Etheric
Fruit Size
Sul
Sal
Sul
Physical
Sul
Back Brain
Cl
Cosmic Substance
Earthly Forces
Na
Stomach
Pointed Leaves
Sn
506
Rhythmic
Hg
503
Round Leaves
Sul Mer
S
H
Mg
Sal Mer
Mid Brain
Small Intestine
Cosmic Forces
Earthly Substance
Tap Roots
Pb
507
Sul
Al
Sal
Ag
505
Ramified Roots
Sal
P
501
Sal
Fore Brain
Si
Large Intestine
Nerve Sense
Sal
Energetic Bodies, Hydrogen, Magnetism, Steiner Physical Forces,
Physical Body Chemistry, Alchemy, Planets, Metals, BD Preps,

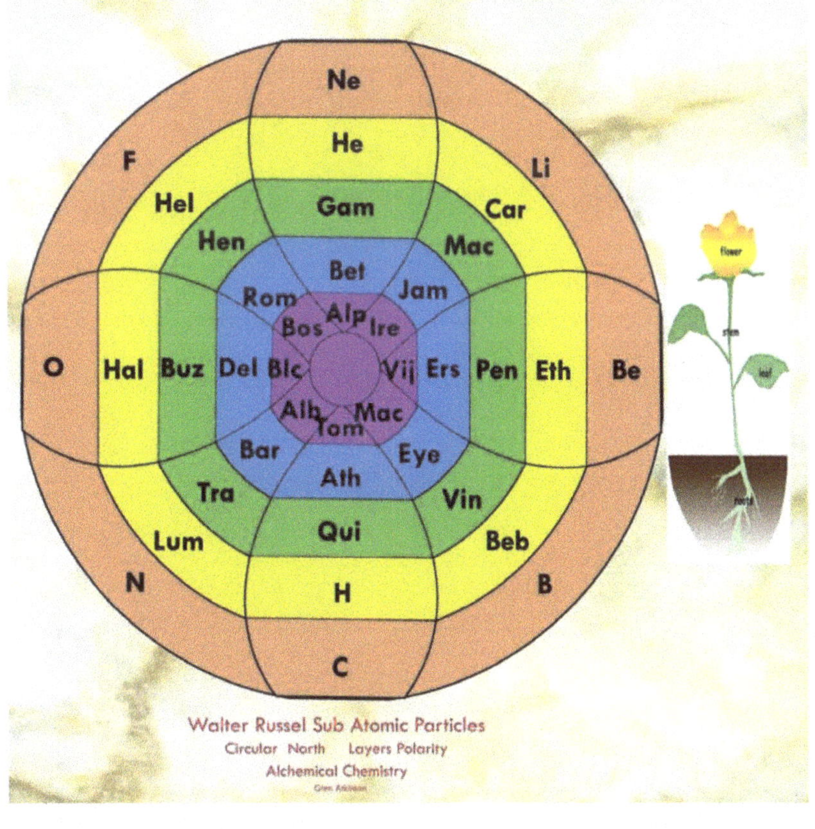
Ne
He
F
Li
Hel
Gam
Car
Hen
Mac
Bet
Rom
Jam
Alp
Bos
Ire
O
Hal
Buz
Del
Blc
Vij
Ers
Pen
Eth
Be
Alb
Mac
Tom
Bar
Eye
Ath
Tra
Vin
Qui
Lum
Beb
N
B
H
C
Walter Russel Sub Atomic Particles
Circular North Layers Polarity
Alchemical Chemistry

The Planets

3 x 4

3 Modes of Action

4 States of Psychology

12 FOLD PLANETS

	THINKING	*FEELING*	*WILLING*
PHYSICAL-**primary** ***Personal*** ***Unconscious***	☿ ♊ **Mercury** Comunication,Opinions, Ideas Local travel, Letters, Telephones 3rd & 6th houses Hermes Astral 14-21	☽ ♋ **Moon** Nurture, Feelings Emotions Mother Home Family 4th house Hera -Demeter Etheric 7-14	♂ ♈ **Mars** Impulse to Action, Anger Assertiveness. Sex 1st & 8th house Ares Physical 0-7
SOUL - **secondary** ***Personal*** ***Conscious***	♃ ♐ **Jupiter** Understanding, Wisdom Morals Extravagance, Indulgence, Big Long distance travel Ideals 9th house Zues Sentient Soul 21-28	♀ ♉ **Venus** Facilitation, Love, Romance Artistic expression, Friends Socialising,.Relationships 2nd & 7th houses Aphrodite Intellectual Soul 28-35	♄ ♑ **Saturn** Manifestation, Rules, Authority, Father, Order Restriction, Contraction 10th & 11th house Cronus Conscious Soul 35-42
COLLECTIVE UNCONSCIOUS **tertiary**	♅ ♒ **Uranus** Genius, Inspiration, Occultist Individuality, Uniqueness Eccentricity, Impulsiveness 11th house Oranus Spirit Self 42-49	♆ ♓ **Neptune** Imagination, Fantasy Mystic Dreams, Illusion, Enchantment Compassion, Psychic 12th house Poseidon Life Spirit 49-56	♇ ♏ **Pluto** Intuition, Shaman, Trust Power, Passion,Obcession Transformation, Change 8th house Hades Spirit Man 56-63

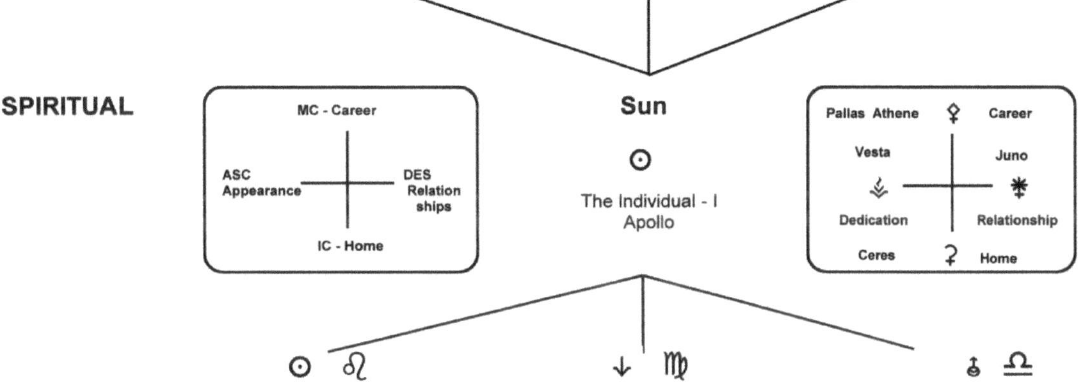

SPIRITUAL ***Collective*** ***Conscious***	☉ ♌ **Sun** "The Word" Apollo "Atmosphere" 77- 84	♍ **Vulcan** Manifestation from spirit Hephaestus "Region of Oceans & Rivers" 70-77	♎ **Transpluto** Enlightenment Persephone "Solid Land" 63-70

Aspects:

☍ □ ∠ ⚼ ⚻

Tensions Opposition, Square, Semi square, Sesquiquatrate:- **Challenges**

Easy : - Trine, Sextile, Semi sextile, Quintile ; Conjunction **Works well with**

△ ⚹ ⚺ Q ☌

Use a word from each planet, make a sentence, play with it with your imagination then make another one.

Glen's Medicine Wheel

**Orientate the picture to Magnetic North.
A bottle of water can be placed on any spot
to accumulate the energy of that spot.
This can be sprayed over the area of the 'pest',
or taken as a remedy.**

ANIMAL

A1	Coelenterata
A2	Echinodermata
A3	Tunicata
A4	Mollusca
A5	Vermes
Insects	
A6	Bacteria
A7	Nematodes
A8	Grubs
A9	Flying
A10	Blood
Pisces	
A11	
A12	
A13	
A14	
A15	
Amphibia	
A16	
A17	Newts
A18	Salamander
A19	Frogs
A20	Toads
Reptiles	
A21	
A22	Crocodiles
A23	Turtles
A24	Snakes
A25	Lizards
Birds	
A26	
A27	Flightless
A28	Ducks
A29	Song
A30	Predator
Mammals	
A31	Marsupials
A32	Rodents
A33	Herbivores
A34	Carnivores
A35	Humans

HUMAN

Bi Polar	
H1	
H2	
H3	
H4	
H5	
Schizophrenia	
H6	
H7	
H8	
H9	
H10	
Autism	
H11	
H12	Alienation
H13	
H14	
H15	Catatonic
Hysteria	
H16	
H17	
H18	
H19	
H20	
Psychopathy	
H21	
H22	Egotist
H23	Narcisist
H24	Sociopath
H25	Psychopath

PLANT

Fungi	
P1	
Algae	
P2	
P3	
Mosses	
P4	Hornworts
P5	Liverworts
P6	Bryophyta
Ferns	
P7	Equiseteidae
P8	Lycopodiae
P9	Polypodidae
Gymnosperms	
P10	Gnetiae
P11	Cycadidae
P12	Pinidae
Angiosperms	
P13	Amborellaceae
P14	Magnoliaceae
P15	Liliaceae
P16	Fabaceae
P17	Malcaceae
P18	Asteraceae

MINERAL

M1	Metals
M2	Obsidian
M3	Basalt
M4	Slate
M5	Serpentine
M6	Schist
M7	Granite
M8	Clay
M9	Limestone

**This picture and these
associations are still a
'conjecture'.
So far so good, but there is
still a ways to go to 'prove'
all the placements, so play
around and let me know
what works for you.
garuda@xtra.co.nz**

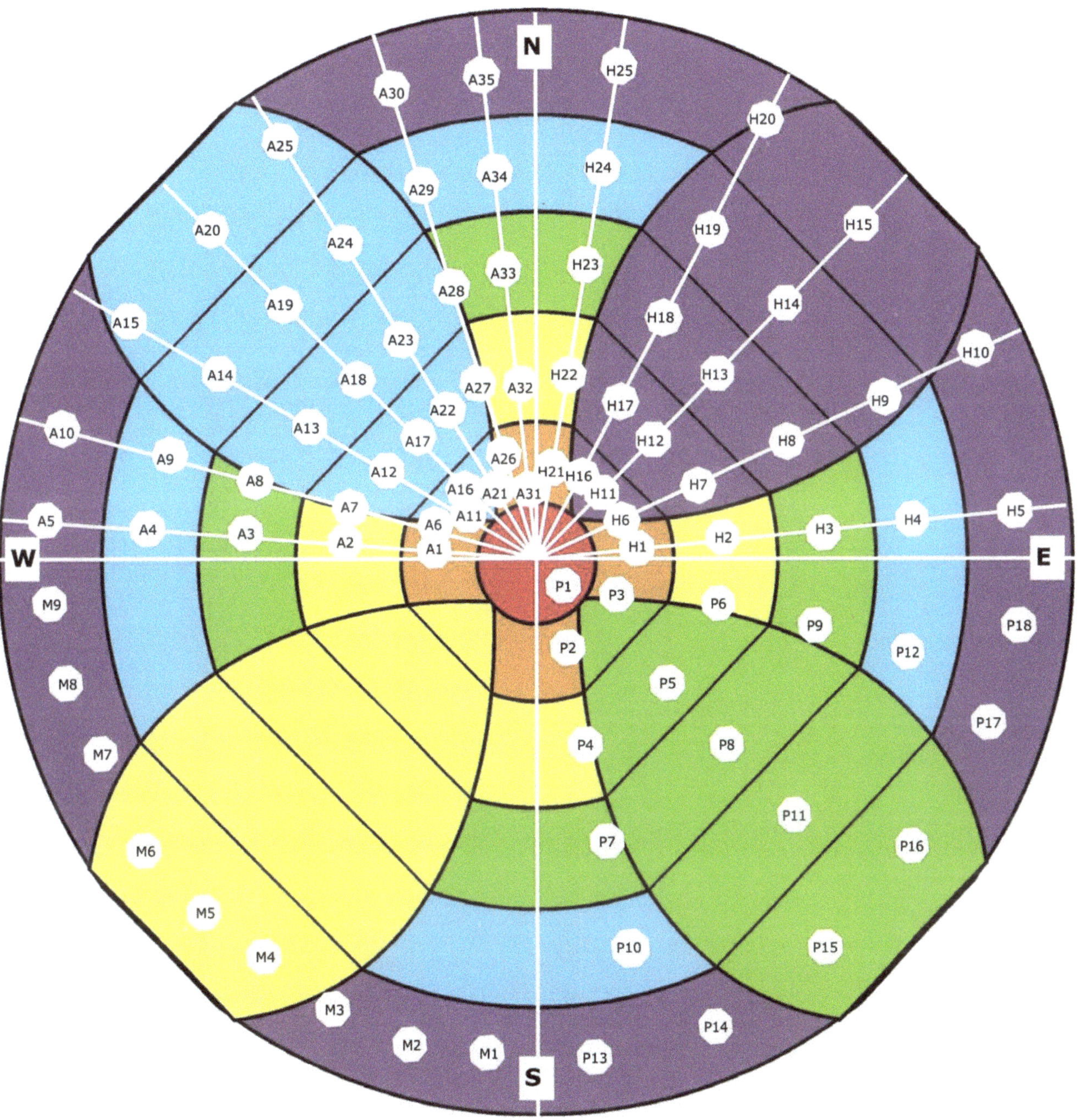

Glens Medicine Wheel

www.ingramcontent.com/pod-product-compliance
Lightning Source LLC
LaVergne TN
LVHW060628110826
845147LV00015B/958

* 9 7 8 0 4 7 3 7 6 5 5 1 4 *